The Calculus Primer

$\infty \ 0 \int \pi \int 0 \ \infty$

Robert J. Madison

authorHOUSE®

AuthorHouse™
1663 Liberty Drive, Suite 200
Bloomington, IN 47403
www.authorhouse.com
Phone: 1-800-839-8640

First published by AuthorHouse 03/07/08

ISBN: 978-1-4343-2407-8 (sc)

Library of Congress Control Number: 2007905703

Printed in the United States of America
Bloomington, Indiana

This book is printed on acid-free paper.

In retirement, after eight years of teaching chemistry and physics in high school, six years of teaching mathematics in college and 16 years of teaching computer science in college, a veteran teacher uniquely presents sixty plus pages of text and illustrations on the origin and purpose of differential and integral calculus. This presentation is regarded by some as a model of simplicity serving as a refresher for former students and as an introductory course for beginners.

Have you ever wondered, "How can I use basic calculus?" Some truly simple examples are given early in the first ten pages including the fine details that you may need. These include an example of the calculation of the instantaneous speed at any point on the parabolic curve, $s = t^2 + 1$, if the distance, s, is depicted on the vertical axis and the time, t, on the horizontal axis.

Also included is the area bounded by the parabolic curve and the closed interval, $0 \leq t \leq 2$, on the horizontal axis. These two magnificent feats were accomplished in the later half of the 17th century by the creators, Isaac Newton and Gottfried Leibniz. These miraculous accomplishments were both based upon the process of differentiation because the process of integration is simply the reverse process to the differentiation process. Both processes were provided with some refinement by the successors to the creators and were combined to become known as the Fundamental Theorem of Calculus.

Many mathematicians labored for over 100 years to place Calculus on a firm mathematical basis. Contrary to traditional mathematical philosophy, Calculus appeared to be based on faith. Rigor was finally accomplished by development of the limit concept which was promoted by a fondling who many years earlier had been left on the doorsteps of the St. Jean Le Rond Catholic Church in Paris (1717). He became known as Jean Le Rond d'Alembert. He was one of the first to realize that an infinite sum of terms could possibly be written as a finite sum even though the number of terms went on endlessly.

The limit concept and the possible production of a finite sum to an infinite number of terms helped to place Calculus on a firm mathematical basis.

iv

Acknowledgements

This book has been conceived over a long period of time. It was not until after retirement that the opportunity arrived to undertake this project. But even so, without the aid and encouragement of many, this book would not have been completed.

Steve Dunn, with his CorelDraw program, made it possible to improve the graphs and drawings originally submitted. My wife, Betty Broaddus Madison, has offered encouragement and proof reading throughout and has not been highly offended by the time commitment to this booklet.

My daughters, Leah M. Cheshire, Kathy M. Wilson and Carol M. Ludden, have been encouraging throughout.

Edgar N. Reyes, Ph.D., Dept. of Mathematics, Southeastern Louisiana University, Hammond, Louisiana, has been quite helpful and has provided two great textbooks on Calculus.

Ellen Lamarque, Mathematics Teacher & Curriculum Specialist, St. Tammany Parish School Board, Covington, LA, has provided some proofreading services and suggestions.

Table of Contents

vi

Preface

If you have never taken a course in calculus, perhaps you should skip the preface. Actually, it is a short review. After approximately the first ten pages, you can return to the preface. Starting with the position equation, $s = \frac{1}{2} at^2$, by the process of differentiation, we can produce a differential equation, called the velocity equation, which has tremendous power. By reversing the differentiation process, we can integrate the velocity equation and return to the original position equation, $s = 1/2\, at^2$, which gives the area under the curve of the velocity equation, $v = at$. Begin with the position equation, $s = 1/2\, at^2$.

$s = \frac{1}{2} at^2$ Multiply the coefficient by the power and reduce the power by one

$ds/dt = at$ After performing the process of differentiation with respect to time,

$v_i = at$ we have the instantaneous velocity which is a ratio of differentials, ds/dt

$ds = at\, dt$ Multiply both sides by dt

$\int ds = \int at\, dt$ Integrate (raise the power by 1, divide by this new power and add a constant)

$s = at^2/2$ From above, the constant term is equal to zero

This book is not intended to be a textbook. Many textbooks will present all the topics normally covered in a full semester course or a two semester course in calculus and analytic geometry. The author believes that most students can't see the forest because of the trees. We intend to cover the forest in this book by dealing with such questions as: "What is the basis of calculus? When do we employ calculus? Can I understand Calculus?"

To begin, we want to make it clear that differential calculus developed because as the 17th century dawned, mathematicians yearned to develop a process that could provide the instantaneous rate of change of position for a moving body. Today, we are well aware that the automobile speedometer approximates this process. But mathematicians did not want an approximation method. The calculation of average speed, $\Delta s/\Delta t$, was no problem. The instantaneous speed, v_i, was quite a problem because it was obvious that the change in distance, Δs, and the change in time, Δt, had to be made as small as possible to approximate the instantaneous speed. But approximations were not good enough. They wanted the process to yield the exact answer. In order to do this, mathematicians had to freeze the changes in both distance and time. But when this is done, motion is destroyed and we get $0/0$ on the left side of the equation. The answer can be any number in the universe but the direct calculation of this value is impossible. Mathematicians say that this value is indeterminate. But the right side of the equation must have the same value as the left side. On page 2 of the text, we show that the v_i, in the example given, has many values depending upon the point under consideration on the graph. If $v_i = 2t$, then at the point, P_1, the instantaneous speed is simply $2 * 1$ or 2 ft/sec. At the point P_2, the instantaneous speed is 4 ft/sec. In other words, we have completely dismissed the problem of dividing zero by zero. We have simply dodged the issue by turning our attention to the right side of the equation. Division of any nonzero number by zero is invalid in mathematics but there is nothing invalid about the expression, $0/0$. It simply cannot be calculated directly. The process of differentiation was accomplished by multiplying the coefficient by the power of the variable, dropping the power by one and dropping the constant. In this manner, mathematicians created the first dodge in the problem of evaluating an indeterminate. This produced a clamor of protest from some mathematicians, with some justification. But the major justification was, "it usually worked". The second great problem, for 17th century mathematicians, was the process of calculating the area between a true curve, the horizontal axis and the upper and lower bounds on this axis. This was an elusive problem but Isaac Newton and Gottfried Leibniz discovered that this could be accomplished by reversing the differentiation process. This second process was referred to as anti-differentiation or integration. This integration process was the beginning of what we now call integral calculus. We simply reverse the process of differentiation. In essence, we multiply the varying heights of an infinite number of rectangles by the zero width of each rectangle (after the limit is taken). The process also sums all of the rectangular areas and presents the exact total of the rectangular areas. The 17th was quite a century. Besides solving the problem of finding the instantaneous velocity at any point on the function curve, the answer was provided to the problem of finding the exact total area between the curve, the horizontal axis and the left and right bounds on this axis. This answer was exact, not an approximation. Can this be?

Well, in this integration process, we have encountered another indeterminate. This is the expression, $\infty * 0$. There is no way to directly calculate this value. So again we must provide a dodge. Here, the dodge is the process of integration, as discovered by Newton and Leibniz. With a polynomial, we raise the power by one, divide by the new power and evaluate the area by inserting the values for the limits of integration. In a nutshell or two, we have summarized the process of differentiation and the reverse process which we call integration.

In the accompanying text, we provide example after example of areas that can easily be calculated algebraically. These exact answers can also be produced by the integral calculus. This is done to promote your confidence in this new branch of mathematics. The new branch of mathematics has two big advantages over algebra. Unlike algebra, the two new processes, generally speaking, have the power to calculate instantaneous speed and exact areas even though one boundary is curved. These processes can also be employed to produce other equations which give us a much better understanding of our environment. We can begin with Galileo's empirical equation, $s = \frac{1}{2} at^2$. Notice that both, the original speed and the original distance, are equal to zero. The symbol, a, represents a constant value for acceleration. We could use the symbol, g. This is the symbol for the constant value called the acceleration of gravity on earth which is 32 ft per sec per sec in English units or 9.8 meter/sec^2 in the metric system. Now, we take the liberty of being partially repetitious:

$s = -1/2\ gt^2$ The minus sign is employed to indicate downward acceleration

$ds/dt = -gt$ Differentiation gives the instantaneous velocity at any time, t (you can believe)

$v_i = -gt$ We will again take the derivative with respect to time (a 2nd derivative)

$dv/dt = -g$ The 2nd derivative gives the value of the negative acceleration

Now we integrate and expect to return to the original equation

$\int dv = \int -g\ dt$ Integrate to obtain the instantaneous rate of change of position

$v_i = -gt$ The formula for instantaneous velocity (constant of integration = 0)

$ds/dt = -gt$ The speed equation is presented in differential ratio form

$ds = -gt\,dt$ Multiply both sides by the differential, dt

$\int ds = \int gt\,dt$ Integrate and return to the position equation

$s = -gt^2/2$ The constant of integration is 0 (from above)

Don't be alarmed if you do not understand all of the preceding text. By providing simple example after simple example and doing all the busy work for you, the learning is as painless as it can be. You will gradually acquire the mathematical maturity needed.

This introductory presentation of calculus is concluded by the Appendices. These helpful topics are included to allow the student the opportunity to review many of the vital topics that precede any course in calculus.

Literal equations, for example, are a much neglected topic in most algebra courses but no topic is of greater importance in applied mathematics. If you don't use it, you lose it.

The five basic right triangles and the ratios of these sides are basic in trigonometry and calculus. We must be able to recall the essentials of both base 10 and base e logarithms. A review of the exponential functions may be necessary. A short summary of the inverse functions may also be needed and we must consider a refresher on the linear elements such as the point-slope form of a line.

Calculus deals with equations and their corresponding functions. The quadratic formula is certainly a necessity and the imaginary roots need to be mentioned since most graphs do not display roots on the complex plane.

Included in the appendices are nine graphs which you should be able to recall by name. Of course, we do not claim to cover all topics in both Differential and Integral Calculus. Our booklet was prepared in order to simplify the presentation of calculus. There is no reason for an intelligent layman to be ignorant of the reason for the development of calculus. In essence, any scientific society must be able to calculate the rates of change in our environment. Without this capability, we remain a limited society. We also have great need of a method for area calculations when one boundary is curved (not a straight line).

We also need to establish the fact that calculus provides exact answers, not approximations. Of course, the answers provided may be incomplete due to the limitations of the machines or the nature of the number system. But this is not the fault of the calculus.

Many former students of calculus are still wondering what it was all about. One statement is quite clear. Generally speaking, when rates of change are involved, we depend upon the calculus to provide the correct answers in accordance with the values of the independent variable. When we encounter area problems that involve a curved boundary, we usually depend upon the Calculus to provide the exact area.

x

Short Selected Bibliography

Barnett, Ziegler & Byleen. College Algebra With Trigonometry. Sixth Edition. McGraw-Hill. New York. 1999.

Bell, E. T. Development of Mathematics. Second Edition. McGraw Hill. New York. 1945.

Dantzig, Tobias. Number, The Language of Science. The Macmillan Company. 1954.

Kelley, W. Michael. The Complete Idiot's Guide To Calculus. The Penguin Group. (USA). New York. 2003.

Kelley, W. Michael. Complete Idiot's Guide To Pre-calculus. The Penguin Group (USA). New York. 2005.

Mendelson, Elliot. 3000 Solved Problems in Calculus. McGraw-Hill. New York. 1988

Seife, Charles. Zero. The Biography of a Dangerous Idea. Penguin Books. 2000.

Smith, Robert & Minton, Roland. Calculus. McGraw Hill. New York. 2002.

VerNooy, Stan. Straight Forward Calculus AB. Garlic Press. Eugene, OR. 1994

Zzzzzzzz. Many other references, lost in Hurricane Katrina. 2005.

Identifying the Original Goal

If a racing car is driven 120 miles in 60 minutes at various speeds, what is the average velocity (speed), v_a? With a little common sense, we can produce a formula, such as $v_a = s /$ t, where s is distance and t is time. The dimensions such as feet or miles are of our own choosing. In the example provided above, we have chosen these to be miles and minutes which gives an average velocity (speed), v_a, of 2 miles per minute or 120 miles per hour. But this average speed over the sixty minutes is quite different from the fluctuations registering on the car's speedometer. At any instant of time, the speedometer gives it's version of the instantaneous speed, v_i, which was beyond the power of mathematicians at the dawn of the 17th century.

In the later half of the 17th century, two mathematicians working independently, showed us how to calculate the speed at any split second of time if the relationship between distance and time could be graphed as a function. For example, the formula, $s = t^2$, describes such a relationship. This gives us a well known curve called a parabolic curve. From this curve, the calculation of the <u>instantaneous</u> speed, v_i, <u>at any point</u> on the curve was an astounding achievement at this time. The two great mathematicians were Isaac Newton of England and Gottfried Leibniz of Germany. Many mathematicians were duly impressed but some were highly critical. In regard to the infinitesimals defined, George Berkeley, an Irish Bishop, was especially caustic with remarks such as: "Direct impossibilities and contradictions with quantities that are neither finite nor infinitely small nor yet nothing. May we not call them the ghosts of departed quantities. He who can digest a first , second or third fluxion (differential) need not, methinks, be squeamish about any point in divinity." But differential calculus usually worked and faith would follow. We will return to this topic later. First, let us take a look at the definitions provided in this new and revolutionary branch of mathematics.

x – Usually the independent variable but the choice could be t for time   (on horizontal axis)

y - Usually the dependent variable but the choice could be s for distance (on vertical axis)

Lim – The horizontal leg of a right triangle approaches 0 which is the limit of reducibility
$\Delta x \to 0$

Δx - a finite increment on x, increases the 1st coordinate to begin locating a 2nd point on the curve

Δy - a finite increment on y, increases the 2nd coordinate to complete the location of the 2nd point

Tangent line - line that touches the curve at only one point, at which it has the slope of the curve

Secant line - a line that normally touches the curve at only two points, giving the ratio, $\Delta y/\Delta x$

$\Delta y/\Delta x$ - the ratio of two finite (measurable) increments (this gives the average speed)

dy/dx – the finite ratio of two infinitesimals (two immeasurable quantities – each less than any named value > 0). This is the elusive (long sought) instantaneous speed, v_i, that remains after the limiting process like the smile on the face of Alice's Cheshire cat.

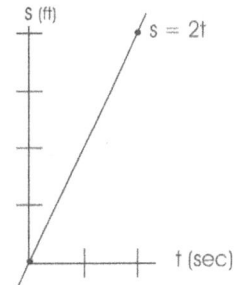

A constant speed always produces a straight line graph.
The graph at the right is a perfect example.

Differentiation, Creation of the Process

If the speed changes, a higher mathematics is needed to give the instantaneous speed at any split second of time. Since the speed is changing, the graph will be truly curved. The following function, $s = t^2 + 1$, is a fine example of such a relationship between the variables distance, s, and time, t. For simplicity, let us assume that the distance is in feet and the time is in seconds. In order to find the instantaneous speed, we employ a differentiation process which is similar to that devised by Newton and Leibniz except that we divide by Δt before taking the limit (to avoid possible division of a non-zero value by zero).

Distance, s, is a function of time, t	$f(t) = t^2 + 1$
Change the function name to s	$s = t^2 + 1$
Increment the two variables	$s + \Delta s = (t + \Delta t)^2 + 1$
Square the binomial on the right side	$s + \Delta s = t^2 + 2t\Delta t + \Delta t^2 + 1$
Subtract s from both sides	$\Delta s = 2t\Delta t + \Delta t^2$ [t^2 and the constant falls out]
Factor Δt out of the binomial on the right	$\Delta s = \Delta t\,(2t + \Delta t)$
Divide both sides by Δt	$\Delta s/\Delta t = 2t + \Delta t$

Take the limit on both sides
  as Δt approaches zero

$$\lim_{\Delta t \to 0} \Delta s/\Delta t = \lim_{\Delta t \to 0} (2t + \Delta t) \quad \text{[taking the limit}$$
after dividing by Δt]

$$0/0 = 2t + 0$$

This process is called differentiation $ds/dt = 2t$

$$v_i = 2t$$

Plotting the graph, we see that at $P_1(1,2)$ $v_i = 2 * 1$ ft/s (instantaneous speed)
  we see that at $P_2(2,5)$ $v_i = 2 * 2 = 4$ ft/s
  we see that at $P_3(3,10)$ $v_i = 2 * 3 = 6$ ft/s
  If t = 3 1/3 seconds $v_i = 2 * 10/3 \approx 6.67$ ft/s

Clearly, the instantaneous speed is 2t. By referring to the graph and plugging the time values into the instantaneous speed equation, $v_i = 2t$, we obtain the instantaneous speed at any selected time.

Berkeley could argue about the left side but 0/0 is legitimate in mathematics and it may represent any speed in the universe which is determined by the graph's point of reference. As an example, the speed at t = 15/2 seconds is: 2(15/2) = 15 ft/sec.

Since a differential is a variable that approaches 0 as a limit, it is impossible to draw a differential but we can construct a ratio, h/w, that has the same value as ds/dt at P_1. Construct a horizontal line of indefinite length through P_1 and drop a perpendicular from any point on the tangent line to provide the intersection which gives the lengths of h and w.

Differentiation, Magnify the Graph

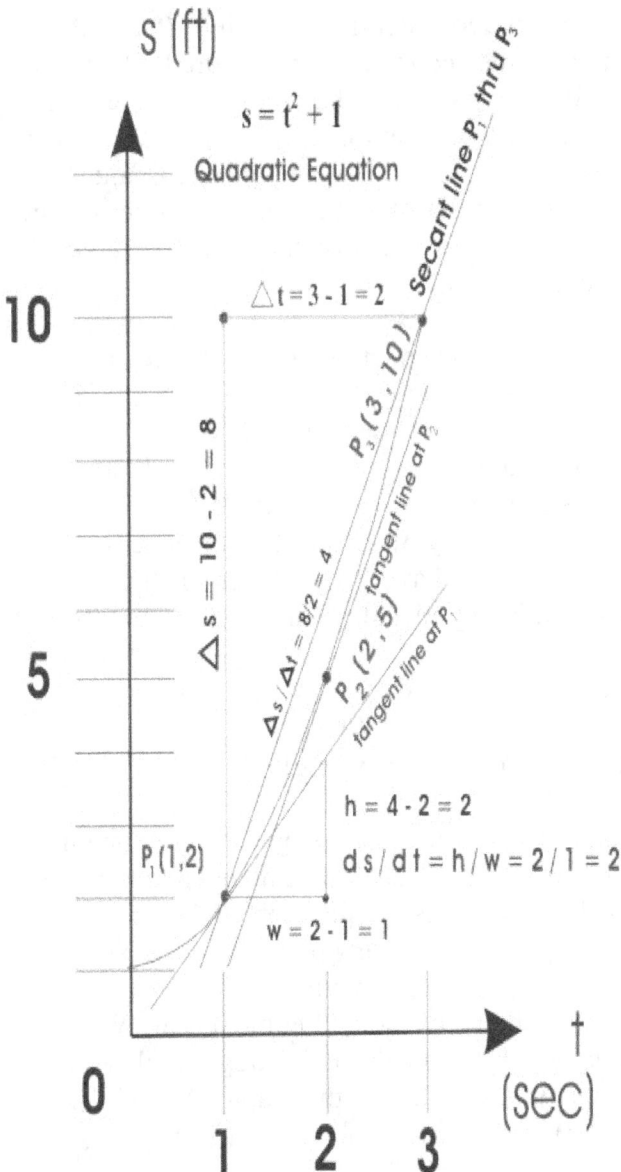

Taking the limit as "delta t" goes to zero appears to be a stumbling block in the understanding of the process of differentiation. The realization that 0/0 is a perfectly valid expression that can equal any number in the universe and the substitution of values for t in the evaluation process should not cause any problem. The symbol, t, is simply a variable that is employed to provide corresponding graph heights as this independent variable changes. Each value of 0/0 results from a corresponding t-value and there exists an infinite number of heights. Refer to the bottom of page 7 for further explanation.

Differentiation - The Stunning Vital Process

Let us graph the parabolic equation, $s = t^2$. Now, we consider the two points, P_1 and P_2. Obviously, the graph passes through these two points and the coordinates are $P_1(1,1)$ and $P_2(2,4)$. Then $\Delta t = 2-1$ and $\Delta s = 4-1$. Therefore, $\Delta v = (4-1)/(2-1)$, implies that the average speed over this interval on the horizontal axis is 3 ft/sec. But this is not the instantaneous speed because the time is finite, 2-1 (seconds). Remember, we must <u>not</u> have a finite time if we want the instantaneous speed, v_i. First, we must let $\Delta t \to 0$. When the time is finite, we have the average speed over an interval of time, v_a, not the instantaneous speed, v_i. To make this point clear, let us start with the equation, $s = t^2$.

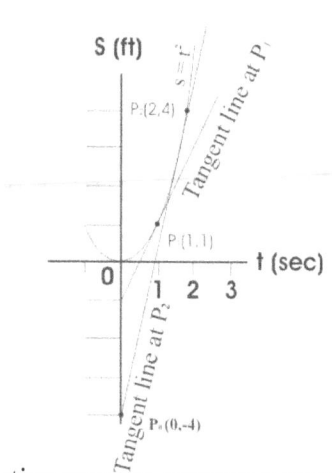

Given the parabolic equation, $s = t^2$ (one unit below the curve, $s = t^2 + 1$)

Increment both sides $\quad s + \Delta s = (t + \Delta t)^2$

Expand the right side $\quad s + \Delta s = t^2 + 2t\Delta t + \Delta t^2$

Subtract s from each side $\quad \Delta s = 2t\Delta t + \Delta t^2$

Divide both sides by $\Delta t \quad \Delta s / \Delta t = 2t + \Delta t$ [we can do this because we have not yet taken the limit]

Take the limit on both sides $\quad \lim_{\Delta t \to 0}(\Delta s/\Delta t) = \lim_{\Delta t \to 0}(2t + \Delta t)$

As dt goes to zero $\quad ds/dt = 2t + 0$

so <u>must</u> ds $\quad 0/0 = 2t$

$\lceil$ both are legs of same rt. triangle $\rceil$

$v_i = 2t$

Notice that the starting height of the parabola has no effect on the slope at any point on the curve. The instantaneous speed, ds/dt, is the slope of the tangent line at any point on the curve. The expression, 0/0, is indeterminate but we dismissed this problem because it has the same value as the right side of the equation, 2t. We pick any point on the curve and simply substitute the value of t at that instant on the curve to obtain the instantaneous speed at this time. Isn't this great? In fact, this is almost unbelievable. Let me repeat: We have v_i at <u>any</u> <u>point</u> <u>on</u> <u>the</u> <u>curve.</u> This is similar to having your cake and eating it too! We simply concentrate on the right side of the equation, $v_i = 2t$, and the graph. At $P_1(1,1)$, the instantaneous speed is $2 * 1 = 2$ ft/s. At $P_2(2,4)$, $v_i = 2 * 2 = 4$ ft/s. At $P_3(3,9)$, $v_i = 2 * 3 = 6$ ft/s.

The concept of differentiation is the most important process in calculus. You should commit this process to memory although you will normally employ the shortcut method when working with polynomials due to the ease of use.

We also receive a big bonus. The essential integration process results from the reversal of the differentiation process as discovered by Newton and Leibniz. They killed two birds with one magnificent stone $\lceil$ a monumental feat $\rceil$.

Galileo's Position Equation

You will recall that the brilliant Italian scientist, Galileo, experimented with balls rolling down an inclined plane and concluded that a vertical tabletop would accelerate a rolling ball at a rate of 32 feet per second for each second of fall at sea level. He called this quantity, g, the acceleration of gravity at sea level. He concluded that the distance fallen was directly proportional to the square of time. Therefore, the distance fallen, s, was equal to some constant times the square of time. This constant had the value, g/2. Since this increase in speed was directed downward, he assigned a negative value to it. Therefore, the experimentally derived formula for a free falling body was, $s = -16\,t^2$. He also proved that contrary to Aristotle, heavier objects did not fall faster than light objects. This was demonstrated by dropping a heavy and a light steel ball from the top of the Leaning Tower of Pisa. In order to properly account for positive and negative quantities, the formula must be slightly modified to $s = s_0 - 16\,t^2$ since the rock starts from an elevated position, s_0, and the acceleration is downward. The parabola is downward opening as displayed on page 8. If a dropped rock takes 3 seconds to strike the water, then we only have one unknown in the constructed equation: $s = s_0 - 16\,t^2$.

Since the final position, s, is zero feet high when the rock strikes the water and the original height is the unknown, s_0, then at impact we have:

$$s = s_0 - 16 * 3^2$$

Add 16 * 9 to both sides $\quad 16 * 9 = s_0$

Rotate 180 degrees $\quad s_0 = 144$ ft

This gives the height of the bridge but to get the instantaneous speed, v_i, after three seconds of free fall, we must differentiate the position function: $\quad s = 144 - 16\,t^2$

Increment the variables	$s + \Delta s = 144 - 16(t + \Delta t)^2$
Expand the right side	$s + \Delta s = 144 - 16(t^2 + 2t\Delta t + \Delta t^2)$
Distribute the 16	$s + \Delta s = 144 - 16t^2 - 32t\Delta t - 16\Delta t^2$
Subtract s from both sides and factor 16	$\Delta s = -32t\Delta t - 16\Delta t^2$
Divide both sides by Δt and factor 16	$\Delta s/\Delta t = -16(2t + \Delta t)$
	$\lim \Delta t \to 0 \qquad \lim \Delta t \to 0$

$$ds/dt = -32t$$

We go from the position function to the velocity function. $\quad v_i = -32t$
Now, let us differentiate this position function (a polynomial) in the easy manner.
This is the short differentiation process with respect to time. $\quad s = 144 - 16t^2$

Drop the constant, multiply the coefficient by the exponent $\quad ds/dt = -32t$
and reduce the exponent by one to get the velocity function. $\quad v_i = -32t$

This is the instantaneous velocity function, v_i, not the average velocity function.
The big advantage of differentiating the long way is the understanding that this brings to the differentiation process. Employ the long method until it is thoroughly understood. If this is not the case, work through these examples again and again until the process is understood and accepted. Remember, downward indicates a negative movement. Upward movement is recognized as positive. Movement to the right is also positive and movement to the left is described as negative.

Reversing the Process of Differentiation

At this point, we have been exposed to the differential equation, $ds/dt = 2t$. From this equation, we can derive other equations, not just numbers. But we need the process of integration which is simply the reverse process to differentiation as shown by Newton and Leibniz.

To integrate; first, multiply the differential equation by the differential, dt

$$ds/dt = 2t$$
$$dt(ds/dt) = 2t\, dt$$

Simplify and move the integral sign into place $\qquad \int ds = 2 \int t\, dt$

Integrate (raise the power by one, divide by $\qquad s = 2t^2/2 + C$
the new power and add a constant) $\qquad s = t^2 + C$

Now, we can reverse the process (differentiate) $\qquad ds/dt = 2t$
(use the exponent as the coefficient, drop the $\qquad v_i = 2t$
power by one and drop the constant)

Yes, we have returned to the original equation. No doubt, the two processes reverse each other and these two shortcut methods that we have employed are great time-savers. Of course, in the process of integration, you may be concerned about the addition of the constant term but this problem disappears when we integrate over an interval. In the introduction to the concept, we establish boundaries for integration on the horizontal axis only but these may be set up on the vertical axis instead. On pages 2 and 5, we detailed the differentiation process to show that the dropping of the constant term is a natural consequence.

We apologize for repetition but the process that follows is so vital that it deserves repetition. Earlier, we had mentioned a free-falling object and the great Italian scientist, Galileo, who experimentally derived the formula, $s = -16t^2$. First, we must slightly modify this equation which does not identify the original height of the object but does imply that the original velocity is zero and the original height is decreasing. Given that the object strikes the water in three seconds, algebraically we arrive at a bridge height of 144 feet. This is the constant that will be dropped in the process of differentiation.

But to find the velocity at which the object strikes the water, we <u>must</u> desert algebra. Beginning with the position function from above, we must obtain the first derivative which gives us the instantaneous velocity, v_i, at <u>any point on the curve.</u> $\qquad s = 144 - 16t^2$

Differentiate $\qquad ds/dt = 0 - 32t \qquad$ [dropping of the constant]
Negative velocity since $\quad v_i = -32(3)$
motion is downward $\quad v_i = -96$ ft/sec

This example clearly presents the value of calculus. Algebra could give us the height of the bridge if we timed the fall, but algebra could not give us the instantaneous velocity at exactly three seconds of free-fall or <u>at any instant</u> we desire. The process of Differentiation was originated by two mathematical giants, Newton and Leibniz. What a magnificent achievement?

Relationships

The formula for the area of a square is $A = lw$. We may see this represented as $y = x^2$. Let us take the 1st derivative of this expression to see if calculus will support this formula for the area of a square. To begin, we will take the 1st derivative. Then we will reverse the process by integrating the first derivative equation. The symbol, y', refers to the 1st Derivative and is called y prime.

$$\begin{aligned} \text{The given function} \quad & y = x^2 \\ \text{Differentiate} \quad & y' = 2x \\ \text{Substitute} \quad & dy/dx = 2x \\ \text{Multiply both sides by dx} \quad & dy = 2x\,dx \\ \text{Now, integrate over the limits, } 0 \le x \le 2 \quad & y = \int_0^2 y'\,dx \\ \text{Substitute} \quad & y = \int_0^2 2x\,dx \\ & y = 2x^2/2 + C]_0^2 \quad \lceil C \text{ is about to drop out} \rceil \\ \text{We substitute the limits and evaluate} \quad & y = 2^2 + C - [\,0^2 + C\,] \quad \lceil +C-C = 0 \rceil \\ & y = 4 \end{aligned}$$

The symbols, y', are called y prime. The symbols, y'', refers to the 2nd Derivative and is called y double prime. The slope at any point on the area function ($y = x^2$) is determined by the process of differentiation. The result is the first derivative equation which is the slope equation, ($y' = 2x$). The slope of the area equation, $y = x^2$, at the x-value, 3, is 6, not 9. Be **careful**! To find the slope of the area equation at any point on the curve, we must substitute the x-value into the slope equation, not the area equation. This example is presented to alert the student to the possibility of substitution into the wrong equation.

The slope equation was obtained by the process of differentiation from the area equation so we know that the slope equation can be integrated. We simply undo that which we just did. After integrating, we substitute the limits within the <u>resulting</u> area equation and evaluate to find the area between the <u>slope equation</u>, the <u>horizontal axis</u> and the closed interval on the horizontal axis.

Change is the most certain aspect of life and the ability to measure the rate of this change is of utmost importance. In our endeavors, we develop functions which are mathematical expressions of our observations. The development of the process of differentiation not only led to our measurement of the rate of change at <u>a point</u> on the graph of a function but to the measurement of the rate of change at <u>any point</u>. That is power!! The process that allows us to develop a formula that we can substitute into to get the slope of the curve <u>at any point</u> on the curve is sneaky! It usually takes a rather large time interval for a student to fully comprehend the significance of this <u>stupendous achievement.</u>

Multiplication is multiple addition and Division is multiple subtraction. When the value of 0/0 is questioned, remember what is being asked. The question being asked, 6/2, is how many times must we subtract 2 from 6 on the number line in order to reach zero. Of course, the answer is three and this is the only answer. We ask the same question of 0/0. Correct answers abound. We may do this any number of times. The correct answer depends upon the value of t.

8

The Free-falling Rock

To find the height of the bridge, we employed algebra. To find the instantaneous speed, we use calculus. We have to employ the trick of making the time interval so short that the speed does not have time to change. In this process, we obtain 0/0 on the left side of the equation but on the right side of the equation we have an expression that we can evaluate. We select the x-value or t-value at which to make this evaluation. This is one of the greatest achievements of the human mind. We did introduce the reverse process to differentiation but did not dwell on the power of integration. Shortly, we will do so and this power will be just as stunning as the power of differentiation.

As stated earlier, to explore the vital facts in regard to the free-falling rock, we needed some help from Galileo who experimentally derived the following position formula. Using Galileo's position function, 1., Isaac Newton, and Gottfried Leibniz, employed the calculus to give us the instantaneous velocity, formula 2., for a free-falling body:

$$1. \quad s = s_0 - 16t^2$$

Multiply the coefficient by the power, drop the power by one, drop the constant

$$ds/dt = -32\,t$$
$$2. \quad v_i = -32\,t$$

Now, this is left for you to do. To determine the height of the bridge, algebra is used which gives the time in seconds required for the rock to hit the water. Next, find the height of the bridge if 4 seconds are required. But algebra cannot provide the instantaneous speed of the rock at the time of impact. Differential calculus must be employed and this provides the instantaneous speed **(slope of the tangent line)** not only at the time of four seconds but at 10/3 of a second or <u>any</u> <u>time</u> <u>value</u> <u>on</u> <u>the</u> <u>curve</u>. The subtle power of this staggers the imagination and don't you forget it.

<u>Graph</u> <u>of</u> <u>the</u> <u>Free-falling</u> <u>Rock</u>

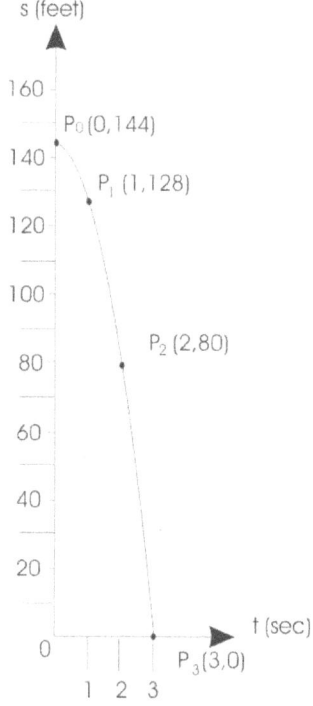

The Limit Concept

At this point, we need to pause and consider a topic that is at the very heart of calculus. This is the concept of limit which was briefly introduced on page 1. Let us employ an example. Perhaps this will somewhat clarify the concept.

What is the final value of 0.999...?

Assign a name to this value　　　　　　　　$N = 0.999...$

Multiply both sides by 10　　　　　　　　$10\,N = 9.999...$

Subtract the first equation from the second　$9\,N = 9$

Divide both sides by 9　　　　　　　　　$N = 1$

Things equal to the same thing are equal to each other.

Therefore, $1 - 0.999... = 0$

In the limit, we can say that $0.999... = 1$

Is it strange that an infinite number of nines to the right of the decimal point will produce a surprising result? Such is the nature of infinity. With infinity, we have accepted a number of astounding statements. Infinity is unchanged by multiplication with a real number. Dividing a real number by infinity yields zero. Infinity is unchanged by the addition of a real number. Is it possible that the multiplying of infinity by zero will yield a real number? We will investigate this result shortly when we employ integral calculus to carry out such a seemingly senseless procedure.

We accept the fact that integration reverses the process of differentiation. In the differentiation process, we multiplied the coefficient by the exponent and dropped the power by one. In integration, the reverse process, we raise the power by one and divide by the new power. We then substitute the limits or add a constant. To reverse a process, we must undo what has been done. On page 1, we attempted to define a differential but didn't tell you that you can multiply and divide by differentials when not in the process of taking the limit. In the process of taking the limit, the differential can take on the value of zero. Of course, the ratio, 0/0, is not illegal in mathematics but it is said to be indeterminate. Once again, perhaps we can provide a dodge that will allow us to determine the value of an indeterminate. We provided a dodge earlier in the process of differentiation. Perhaps, the process of integration would prove to be the process that was needed. Yes, Newton and Leibniz found that integration was the very process that would give us the exact area between the curve, the horizontal axis and within the limits on the horizontal axis.

We now need to look into the logic of the integral calculus. First, let us consider the area of a triangle. On page 10, we see two examples of attempting to fill the interior of the triangle with rectangles. From the diagrams, it is obvious that the greater the number of rectangles employed, the greater the accuracy of the area approximation. Now, if we could simply provide an infinitude of rectangles, then the exact area would result? This is exactly what is accomplished by the process of integration which like differentiation employs the limit concept.

Rectangular Areas

By the formula   Area = ½ ab   a, altitude   b, base
for the area of   = ½ * 12 * 12
a right triangle   = 72 square units

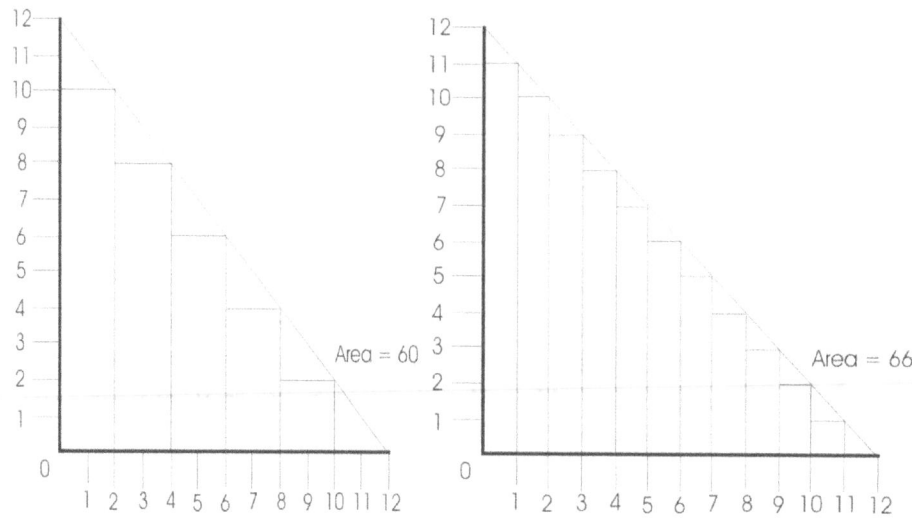

As Δx is diminished, the approximations to the actual area of 72 become more accurate. In the limit then, it appears that we could expect an exact answer. But does this make any sense? How can we possibly add the areas of an infinite number of tiny rectangles with widths of zero? Have faith. The mystical Integral Calculus has this power!

Let us investigate by working with a triangle that has a height of 8 units and a width of 6 units. We employ the algebraic area formula for triangles to determine that the total area is 24 square units. Can the Integral Calculus arrive at this exact answer?

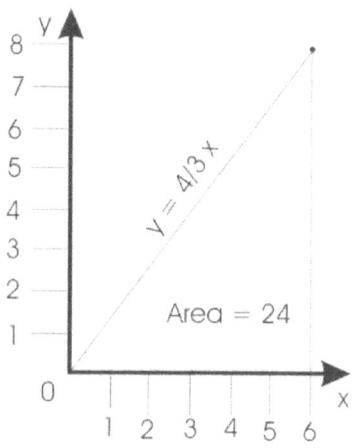

The equation of a straight line through the $y = 4/3\ x + 0$
   origin with a slope of 4/3
Prepare to get the area over the interval   Area = $\int_0^6 y\ dx$
   on the x-axis $0 < x < 6$
Shift the constant to the left of the inte-   = $4/3 \int_0^6 x\ dx$
   gral sign
Integrate (increase power by one and divide   = $4/3\ [x^2 / 2]_0^6$
   by the new power)
Insert the limits of integration   = $4/3\ [6^2/2 - 0/2]$
   and we have agreement   = $4/3\ [18 - 0]$
                              Area = 24   square units

Algebra and Calculus are in agreement, Area = 24

Arc Length - Algebra and Calculus

We can consider the hypotenuse of a right triangle to be an arc length. We can calculate this length by using Algebra as well as Calculus. In this manner, we can check on the validity of the Calculus method. Let us begin with the equation, y = 3x. By an <u>ancient</u> theorem: $s^2 = x^2 + y^2$, $s^2 = 1^2 + 3^2$, $s = \sqrt{1+9}$, $s = \sqrt{10}$. This is the arc length by way of algebra which is the length of the hypotenuse over the interval, $0 \leq x \leq 1$.

The tiny lengths, ds, dx and dy are depicted in the graph below. These are infini- tesimals, smaller than any number that you can possibly name. In a tiny triangle (depicted but actually too small to see), we have: $ds^2 = dx^2 + dy^2$, $ds^2/dx^2 = 1 + dy^2/dx^2$, $ds/dx = \sqrt{1 + (dy/dx)^2}$, $s = \int_0^1 \sqrt{1 + (dy/dx)^2}\, dx$, $s = \int_0^1 \sqrt{1 + 3^2}\, dx$, $= \sqrt{10} \int_0^1 dx$, $s = \sqrt{10}\, x]_0^1$, $s = \sqrt{10}\,(1-0)$, $s = \sqrt{10}$. This is the arc length by way of calculus which is in agreement with the algebraic calculation. Fortunately, we have agreement of Calculus with Algebra. No, this is not luck!

Algebra and calculus produce the same answer in the case of the straight line hypotenuse, but only the Calculus can handle the curved hypotenuse. The symbol, $\int$, implies that we are going to sum an infinite number of arc lengths and the curved hypotenuse may not be any great problem. Such is the power of Calculus and the imagination. When dealing with a straight line graph, dy / dx is always equal to $\Delta y /\Delta x$. This is not so when we work with a truly curved graph. We must take the limit to obtain the ratio of differentials. This is not required for the average velocity, $\Delta y /\Delta x$.

When Δx goes to zero, why do we assign a value of zero to Δy? As Δx shrinks, so must Δy. These are the related legs of the same right triangle. One is some multiple of the other and x times 0 = 0. Don't overlook the indeterminate, 0/0. We must determine one of the many values. By definition of an equation, the left side must be equal to the right side of the equation. In this manner, we determine the values of the indeterminate, 0/0. We substitute any t-value into the independent variable, t, on the right side of the equation. The corresponding ordinate value results.

At the top of page 2 and near the middle of page 4, we went through a lengthy procedure in order to calculate the first derivative of a polynomial which gives us the slope of the tangent line at any point on the curve. Then near the bottom of page 5, we presented a short cut method for this calculation. Simply multiply each term by the exponent, drop the power by one and drop any constant term.

If $y = x^3 + 4x^2 + 5$, then $dy/dx = 3x^2 + 8x$. Now, if you are working with a polynomial, isn't that an easy way to get the slope of the tangent line at any point on the curve?

Areas Under Polynomial and Exponential Curves

Let us consider nearly equal areas under two different curves. The linear equation, $y = 3x + 1$, cuts the y-axis at $y = 1$. The exponential equation, $y = e^x$, does the same. The base, e, approximates the value, 2.718. From the graph, it is clear that the area bounded by the curve, $y = 3x + 1$, the x-axis, and the limits, $0 \leq x \leq 2$, slightly exceeds that under the exponential curve, $y = e^x$, with the same bounds. We superimpose one curve on the other to compare.

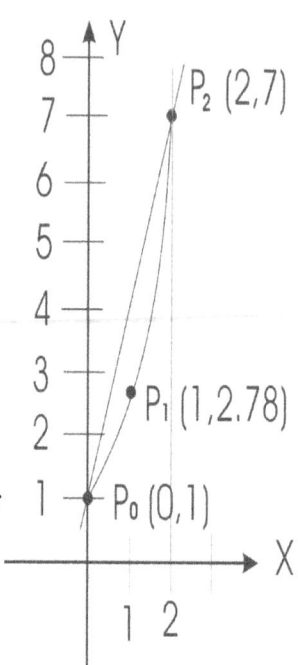

From the graph, we see a right triangle atop two square units that sit below the base of the triangle. By the formula, Area = ½ ab, the area of the triangle is six square units. Adding the two square units, we get a total of eight square units above the x-axis. Employing calculus, we get the same answer: Area $= \int_0^2 (3x+1)dx = (3x^2/2 + x)\rceil_0^2 = 3(2)^2/2 + 2 - 0 = 8$ square units.

But when working with the curved hypotenuse associated with e^x, we find that we must rely on calculus to give us the correct answer. Algebra cannot accomplish this. By calculus: Area $= \int_0^2 e^x\,dx = e^x\rceil_0^2 \approx 2.718^2 - e^0 \approx 7.388 - 1 \approx 6.388$. We have limited the answer to four significant digits. This answer appears to satisfy our expectations as it does not exceed the 8 square units under the hypotenuse (a straight line).

Some students do not understand that calculus gives answers as exact as we need. Of course, no printer can print out all of the digits in an irrational number. Neither can any machine calculate all the digits in this irrational number. But none of this is caused by calculus.

Engineers simply determine the accuracy desired and this is requested. The point that we want to make is that calculus is not an approximation tool. The limit concept does not breed approximations.

I suppose that the above paragraph is a shocking statement but we must eliminate any thoughts about calculus being an approximation process. It is an exact process but calculus cannot change the structure of the real number system.

Charles Seife

Charles Seife

As your tangent approximations get better and better, $\Delta s/\Delta t$ becomes a ratio of differentials, ds/dt, and takes on the value, 0/0, which can equal any number in the universe.

The limit concept places calculus on a firm mathematical basis. If we take the limit on the right side of the equation, we must take the limit on the left side of the equation. All stand-alone differentials become zeroes and the ratio, ds/dt, becomes 0/0, which takes on the same value as the right side of the equation.

In the example presented below, it follows that at P_1, the value of 0/0 is 3 and at P_2, the value of 0/0 is 12. These two values are two elements in the set of possible numbers in the universe.

Begin with the function, $s = t^3 + 2$.

Multiply the term by the power, reduce the power by one and drop the constant

$$s = t^3 + 2$$
$$ds/dt = 3t^2$$
If $t = 1$ then $ds/dt = 3$
If $t = 2$ then $ds/dt = 12$

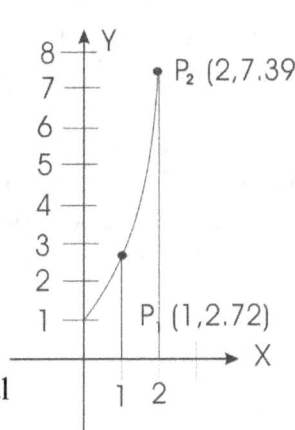

Tobias Dantzig

How can the flat and the straight and the uniform be adapted to the skew and the curved and the non-uniform? Not by a finite number of steps certainly. This can be accomplished only by that miracle-maker, the infinite. The miracle is that it works.

Given a function that produces a true curve, how do we find the length of an arc over any interval? Certainly not by algebra. We must employ a branch of mathematics that can adequately handle some of the tribulations associated with the infinite. We must be able to sum an infinite number of tiny arc lengths or an infinite number of tiny areas that approach zero as a limit. Since we have already summed arc lengths

for the function, $y = e^x$, let us now sum the tiny areas. To do this, we must integrate the function which is quite easy to do for this function. Let us do so over the closed interval, $1 \le x \le 2$.

$$y = e^x$$
$$\text{Area} = \int_1^2 y \, dx$$

$$= \int_1^2 e^x \, dx$$

$$= e^x]_1^2$$
$$\approx 2.718^2 - 2.718^1$$
$$\approx 7.388 - 2.718$$
$$\approx 4.67$$

Area of the trapezoid at the right
$$A \approx a (7.39 + 2.72)/2$$
$$A \approx (2-1)(10.11)/2$$
$$A \approx 5.06$$
the area under the curve, $y=e^x$, is slightly less than that under the the trapezoid over the closed interval $1 \le x \le 2$ as expected

14

The Miracle Maker

Tobias Dantzig

The importance of infinite processes for the exigencies of technical life can hardly be overemphasized. Practically all applications of arithmetic to mechanics, physics and even statistics involve these processes directly or indirectly. Banish infinite processes, and all mathematics pure and applied is reduced to the state in which it was known to the pre-Pythagoreans.

We get the exact length only when each diagonal or horizontal length, the infinitesimal ds or dx, goes to zero. When the number of infinitesimals become infinite, we sum these over an interval to get a finite value over a finite period of time. Previously, we summed an infinite number of the areas in this manner.

Of course, the area formula is quite different from the length formula. In the length formula, we sum the infinitesimals, ds. In the area formula, we sum the infinitesimal products, ydx, which gives us the total area. Let us calculate an area by the calculus method over the closed interval, $1 \le x \le 3$, and by the trapezoidal algebraic method. This line has a slope of 3 and a y-intercept of 2.

It's equation is:

$$y = 3x + 2$$

Area $= \int_1^3 y\, dx$

$\quad = \int_1^3 (3x + 2)dx$

$\quad = 3x^2/2 + 2x]_1^3$

$\quad = 3*3^2/2 + 2*3 - (3*1/2 + 2*1)$

$\quad = 27/2 + 2*3 - (3/2 + 2)$

$\quad = (27 + 12)/2 - 7/2$

Area $= 32/2 = 16$

The area of the trapezoid at the right

$A = a(5 + 11)/2$

$A = (3-1)(16)/2$

$A = 16$

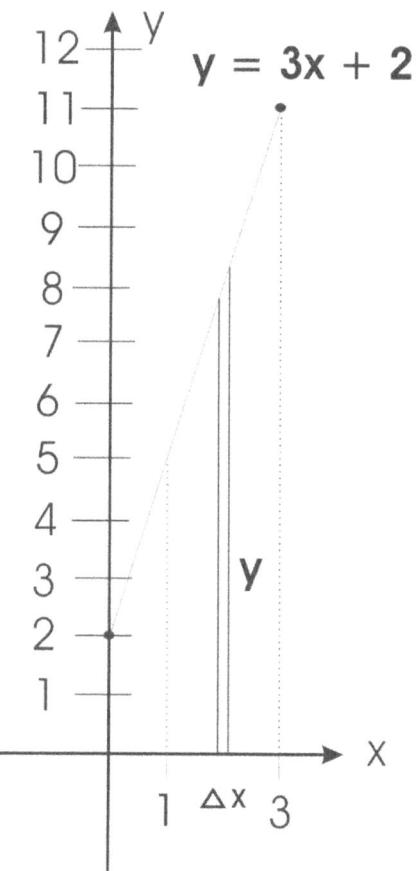

Hopefully, you now have a full realization of accomplishing a miracle. By the process of integration, we have summed an infinite number of zero areas over a finite period of time to obtain a finite area.

Integration of complex curves can lead to involved calculations but you have now been introduced to the deep philosophical implications of a very powerful tool.

Tobias Dantzig

The veritable orgy which followed the introduction of infinitesimals was but a natural reaction. Intuition had too long been imprisoned by the severe logic of the Greeks. Now it broke loose and there was no Euclid to keep its romantic flight in check.

Is it not remarkable then, that in spite of all the loose reasoning, all the vague notions and unwarranted generalizations, so few errors had been committed? "God ahead, faith will follow" were the encouraging words with which D'Alembert kept reinforcing the courage of the doubters. They did forge ahead, guided in their wanderings by a sort of implicit faith in the validity of infinite processes.

The two great mathematicians, were Newton and Leibniz. They encountered two obstacles in their differentiation process. These were the confusion over the indeterminate term, 0/0, and the question of the term, $(\Delta x)^2$, assuming that it had trivial value. It so happened that the area problem was solved by simply reversing the process of differentiation.

If we want to find the total area under the first derivative curve, we first differentiate a given equation to insure that we have an equation that we can integrate (first derivative equation). We can then integrate with confidence and plug in the given limits of integration. This gives us the total area under the curve being integrated.

The given equation $\qquad s = t^3 + 2$

Differentiate $\qquad ds/dt = 3t^2$

Since this is a 1ˢᵗ derivative, we know it is integratable $\qquad s' = 3t^2$

At the right, we present the area formula for any integratable function, s' $\quad$ Area $= \int s'\, dt$

Substitute the expression on the right side, $3t^2$ $\qquad = \int 3t^2\, dt$

Integrate (raise the power by 1, divide by it and add a constant) $\qquad = 3t^3/3 + C$

Simplify $\qquad$ Area $= t^3 + C$

The column at the left is employed for our comments. When integration is over an interval on the x-axis such as $1 \le x \le 2$, the constant, C, is both **added and subtracted** so it falls out of the picture. Consider the equation, $s = t^3 + 2t^2 + 3t$, over the closed interval, [1,2]. We use the area formula in Calculus.

The area under the curve is: Area = s dt over the given interval

$$\text{Area} = \int_1^2 (t^3 + 2t^2 s + 3t)\, dt$$
$$= t^4/4 + 2t^3/3 + 3t^2/2 + C]^2_1$$
$$= 2^4/4 + 2(2)^3/3 + 3(2)^2/2 + C - (1/4 + 2/3 + 3/2 + C)$$
$$= 4 + 16/3 + 6 - (3/12 + 8/12 + 9/6)$$
$$= 12/3 + 16/3 + 18/3 - (3/12 + 8/12 + 18/12)$$
$$= 48/12 + 64/12 + 72/12 - (29/12)$$
$$= 184/12 - 29/12$$
$$= 155/12$$

The area presented is a rational number which is non-terminating. This is the desired form because the computer cannot print out an exact value in the form of all decimal fractions, but it can in the form of a rational fraction.

16

Exercises in Differentiation and Integration

Earlier, we had discussed integration. We said that this process would give the area between the curve and the horizontal axis that is bounded at the left and right by an interval on the horizontal axis. Also, in this discussion, we indicated that this process of integration included taking the limit. It appears that this is similar to multiplying an infinitude of various rectangular heights by zero widths and adding them together. Does this make any sense?

Previously, we considered an indeterminate form, 0/0. But there are other indeterminate forms such as infinity divided by infinity and infinity multiplied by zero. By the process of integration, we multiply an infinite number of various heights by zero widths and sum all of these tiny rectangular areas. This can be done by the process of integration which is the reverse of the process of differentiation. This means that the limit is involved in this process also. So once again, we provide a dodge to the process of evaluating an indeterminate. This time, we simply integrate over the provided limits. These limits tell us where we begin integrating and where we stop integrating.

The power formula for integration is the inverse of the power formula for differentiation. To integrate we increase the power by one and divide by the new power. If no interval is given, we must add a constant of integration. You will recall that we chose to calculate the area that lies below the curve but above the horizontal axis. Yet, the area is bounded also by the upper and lower limits on the horizontal axis.

Let us differentiate the following function. We will follow this with the integration process over the closed interval, $2 \leq x \leq 4$.

$$y = 2x^3 + 5x$$
$$dy/dx = 6x^2 + 5$$

This gives the instantaneous rate of change $\quad v_i = 6x^2 + 5$

Multiply both sides by dx $\quad dy = [6x^2 + 5]\,dx$

Now, integrate over the interval $\quad \int dy = \int_2^4 [6x^2 + 5]\,dx$

$$y = 6x^3/3 + 5x + C\,]_2^4$$

notice the + C and the - C cancel $\quad y = 2(4)^3 + 5(4) + C - [2(2)^3 + 5(2) + C]$

$$y = 128 + 20 + C - [16 + 10 + C]$$
$$y = 148 + C - [26 + C]$$
$$y = 122$$

Notice that the constant, C, simply drops out of the equation.

Now, your assignment is to create two examples similar to the above and work through them. Don't say that you can't because you have been trained to do these things.

Trigonometric Derivatives

Consider memorizing the Trigonometric derivatives. Some are quite difficult to derive and the memorization is not difficult for the basic functions. First of all, simply remember that for one-half of the twelve basic functions, the derivatives of the co-functions are all negative.

Where y' is shorthand for the expression: dy/dx y' is called y prime

$y = \sin x$	$y' = \cos x$	$y = \arcsin x$	$y' = 1 / \sqrt{1 - x^2}$
$y = \cos x$	$y' = -\sin x$	$y = \arccos x$	$y' = -1 / \sqrt{1 - x^2}$
$y = \tan x$	$y' = \sec^2 x$	$y = \arctan x$	$y' = 1 / (1 + x^2)$
$y = \cot x$	$y' = -\csc^2 x$	$y = \text{arccot } x$	$y' = -1 / (1 + x^2)$
$y = \sec x$	$y' = \sec x \tan x$	$y = \text{arcsec } x$	$y' = 1 / (|x| \sqrt{x^2 - 1})$
$y = \csc x$	$y' = -\csc x \cot x$	$y = \text{arccsc } x$	$y' = -1 / (|x| \sqrt{x^2 - 1})$

Note that the derivative of sin x is cos x and, as we know, this derivative gives the slope of the tangent line at any point on the sine curve. If we set the derivative equal to zero, we can find the equations of the horizontal tangent lines to the sine curve. By referring to the graphs in Appendix B or page 21, we see that over the angle range, $0 \le x \le 2\pi$, the only two angles that qualify are $\pi/2$ and $3\pi/2$. In other words, x is the angle at which cos x equals zero. Refer to appendix B for the graph of both the sine and the cosine functions. When we set cos x equal to zero, we can visually observe that the angles giving a zero slope on the sine curve are $\pi/2$ and $3\pi/2$. By definition, we have critical numbers at these two angles on the sine curve. We can see extrema (a maximum and a minimum) at these critical numbers.

Looking at the cosine curve (Appendix B or page 21), we see that we have a minimum at π radians (180 degrees) and maxima at 0 radians and 2π radians. By taking the first derivative, we find the slope of the tangent line at any point on the original curve (cosine curve). Assign a zero slope to the first derivative and we have 0 = -sin x. On the cosine curve, the slope is zero at 0 radians, π radians and 2π radians. At π radians, we see a minimum. At 0 radians and 2π radians, maxima appear.

The maxima and minima that we identified can be seen on the graphs located on Appendix B. Let us now consider the problems with polynomials for which we do not have a graph for a check. First, create the polynomial for the volume of a box without a top. To accomplish this, we will cut out four square corners from a given sheet of cardboard that measures 8 inches long by 3 inches wide. After cutting out the corners and folding, the box length is 8-2x inches, the width is 3-2x inches and the height is x inches. Of course, the formula for the volume of the open box is: Volume = lwh

$V = (8-2x)(3-2x)x$
$V = 4x^3 - 22x^2 + 24x$
$V' = 12x^2 - 44x + 24$
$0 = 12x^2 - 44x + 24$

3 inches width
8 inches length

3-2x inches width
8-2x inches length

Tangent Lines

Employ the quadratic formula to find the critical numbers for zero slopes of the tangent lines to the volume equation above. Substitute into the quadratic formula to find the roots where the slope of the tangent lines to the Volume equation are equal to zero.

$$x = (-b \pm \sqrt{b^2 - 4ac}) / 2a$$
$$x = [44 \pm \sqrt{(-44)^2 - 4(12)(24)}] / [2(12)]$$
$$x = (44 \pm \sqrt{1936 - 1152}) / 24$$
$$x = (44 \pm \sqrt{784}) / 24$$
$$x = (44 \pm 28) / 24$$
$$x = 3 \text{ or } 2/3 \text{ inches}$$

So the original polynomial has slopes of zero at x = 3 or x = 2/3 but x = 3 is impossible because the width is (3-2x) which would give a negative width. The other critical number is 2/3. Use the Second Derivative Test to determine whether this critical number produces a maximum or a minimum. If the result equals zero, we must use the First Derivative Test (If 2nd Derivative Test fails).

From left to right the slope is increasing and f ''(x) > 0 (+)⇒ concave up

f'(x) = 0

From left to right the slope is decreasing and f ''(x) < 0 (-) implies concave down

$$V' = 12 x^2 - 44x + 24$$
$$V'' = 24 x - 44$$
$$= 24(2/3) - 44$$

f ''(x) < 0, (-) ⇒ concave down   f '(x) = 0

At the critical number, x = 2/3, V'' = – 28. According to the 2nd Derivative Test, a negative value indicating a maximum is produced at the critical number, x = 2/3. If the value had been positive at this critical number, then the extremum would have been an impossible minimum. Now we know how to design the box so that from a given amount of material (a rectangular cardboard sheet), we can maximize the storage capacity. Calculus can give us this valuable answer. The bottom of the box has an area of (8-2x)(3-2x) which is about 6.67 times 1.67 giving nearly 11.14 inches^2. We find the volume by multiplying this area by .67 inches, the box height, giving approximately 7.46 inches^3.

The above problem is but one of many problems of a similar nature. Employ the following steps when confronted with such a problem:

1. Construct a continuous function over an interval to be maximized or minimized.
2. Take the first derivative of the resulting (or given) function.
3. Assign a value of zero to the first derivative (to get x-values that produce a slope of zero).
4. Factor the first derivative or employ the Quadratic formula or complete the square to get the roots at which the slopes of the tangent lines are equal to zero.
5. These are the critical numbers which must be tested to determine the extrema, if any.
6. Employ the First or Second Derivative Test to determine whether each critical number of a continuous function is a maximum or a minimum or a point of inflection. At a point of inflection, the rate of change of the velocity function, f'(x), reverses signs. To determine this, we take the derivative of the function, f'(x), to obtain f''(x) which is the acceleration function.

Projectile Motion

A cannonball is fired from a 10 meter high building at a 60 degree angle from the horizontal with a with a muzzle velocity of 100 meters per second. Determine the height that the cannonball will attained. From Appendix B, we find that the sine of 60 degrees, or $\pi/3$. radians is $\sqrt{3}/2$. Multiplying this value by the muzzle velocity will give us the upward velocity of the cannonball. The equation for projectile motion is given below. The constant, g, is called the acceleration of gravity. This has the value of 9.8 meters per second per second which may be expressed as 9.8 meters per second squared. The original height, s_0, is 10 meters high. The original vertical velocity, v_0, is equal to 100 times the sine of $\pi/3$ radians. The projectile will continue to rise until the slope of the tangent line is equal to zero. Setting the first derivative equal to zero will allow us to solve the slope equation for the time required to reach maximum height.

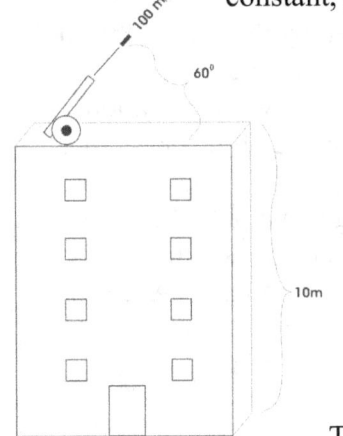

Substitute values into the literal position equation

$$s = -1/2gt^2 + v_0t + s_0$$
$$s = -1/2(9.8) t^2 + 10\sqrt{3}/2\ t + 10$$

The simplified position equation

$$s = -4.9\ t^2 + 86.6\ t + 10$$

The first derivative $\qquad ds/dt = -9.8\ t\ +\ 86.6$

Set first derivative equal to 0 $\qquad 0 = -9.8\ t + 86.6$

Solve for the value of t $\qquad t = 86.6\ /\ 9.8$

$$t = 8.8\ sec$$

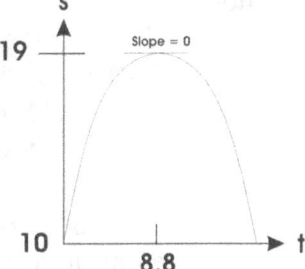

We need to plug the value of t into the position function in order to find the maximum height risen by the cannonball.

$$s = -4.9\ t^2 + 89.6\ t + 10$$
$$s = -4.9\ (8.8)^2 + 89.6\ (8.8) + 10$$
$$s = -379.5 + 788.5 + 10$$
$$s = 419\ meters$$

Now, let's employ the Second Derivative Test. If the result is negative, this implies that the slope is decreasing so a maximum has been found. Let us begin with the First Derivative.

$$v_i = -9.8\ t + 86.6$$

Differentiate to get the 2nd derivative $\qquad dv/dt = -9.8 \quad$ neg., implies a <u>max</u>.
which gives the acceleration (rate of change of velocity)
In order to determine the time during which the cannonball rises, we must get the first derivative and set it equal to zero because the cannonball will continue to rise until the slope of the tangent line is equal to zero.

Actually Firing A Mortar

Actually, firing a mortar from the top of a building may be quite damaging to the structure of the building. Firing from the top of a hill with a known elevation would be more practical and less damaging from a structural standpoint.

The student must realize that the force of gravity does not retard the horizontal speed of the projectile. It is the vertical component of velocity that is retarded by gravity. Since the projectile was fired at an angle, we must multiply the muzzle speed by the sine of the angle to obtain the true vertical speed of the projectile. The vertical speed of the projectile is opposed by the component of the gravitational force that gives a negative acceleration. The cosine of the angle is applied to calculate the true horizontal speed of the projectile.

If the muzzle speed had been 150 meters per second and the firing angle only thirty degrees above the horizontal then our first step is to convert the angle from degrees into radians. By referring to Appendix B, we obtain the angle measurement of $\pi/6$ radians. According to our reference, the sine of this angle is ½. One-half times 150 meters/sec gives us a vertical muzzle speed of 75 meters per second that is directly opposed by the negative acceleration of gravity which is -9.8 meters/sec^2.

The calculation of the time required to reach maximum height begins with the position equation.

$$
\begin{aligned}
&\text{The position equation} && s = -1/2\,gt^2 + v_0 t + s_0 \\
&\text{Differentiate to get velocity equation} && s = -1/2 * 9.8\,t^2 + 150 * \tfrac{1}{2}\,t + 10 \\
&\text{First, Simplify} && s = -4.9\,t^2 + 75\,t + 10 \\
&\text{the velocity equation} && ds/dt = -9.8\,t + 75 \\
&\text{Set slope = 0} && 0 = -9.8\,t + 75 \\
&\text{Solve for the value of } t && 9.8\,t = 75 \Rightarrow t = 75 \text{ meters/sec} / 9.8 \text{ m/sec}^2 \\
&\text{the time to reach maximum height} && t = 7.65 \text{ sec}
\end{aligned}
$$

We must plug this time value into the position equation to find the maximum height

$$
\begin{aligned}
&\text{Employ the position equation} && s = -\tfrac{1}{2}\,gt^2 + v_0 t + s_0 \\
&&& s = -4.9 * (7.65)^2 + 75 * 7.65 + 10 \\
&&& s = -4.9 * 58.52 + 573.75 + 10 \\
&&& s = -286.75 + 583.75 \\
&&& s = 297.00 \text{ meters}
\end{aligned}
$$

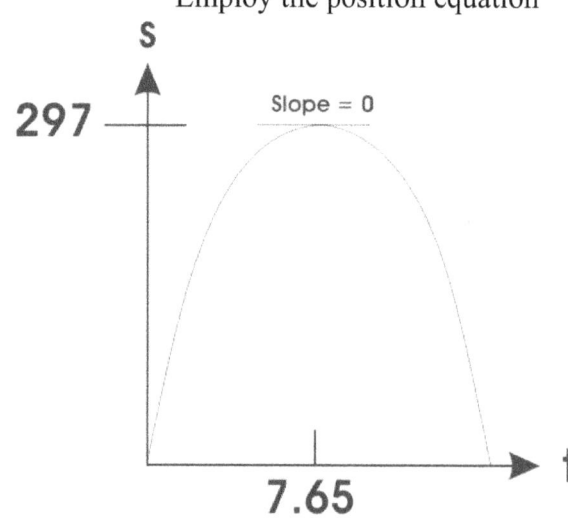

S

297 — Slope = 0

7.65

t

Finding the Area under the Sine Curve

Let us carefully study the sine curve below. We desire to find the area between the sine curve and the horizontal axis. First, let us make an estimate of this area which is obviously less than the area bounded by a height of 1 unit, the t-axis and the limits, 0 and π. The area described is a rectangle which has the area: A = height * width.

A = 1 * π $\approx$ 3.1416 square units. From the graph, we would have guessed that the area to be excluded must be about 1/3 of the rectangle described. This estimate leads us to perform the following guesstimate: Area between the sine curve, the t-axis and the limits, $0 \le t \le \pi$, is approximately 2.095 square units (2/3 * 1 * 3.1416).

Let us compare results. We will employ integral calculus to calculate the exact area under the sine curve over the closed interval, $0 \le t \le \pi$. From an Integration table, we have: $\int \sin x \, dx = -\cos x + c$ and employing the fundamental theorem of calculus, we have the following:
$$\int_0^\pi \sin t \, dt = -\cos t + c]_0^\pi \quad \text{and} \quad F(\pi) - F(0) = -\cos \pi - (-\cos 0) = -(-1)-(-1) = 2.$$
Since we are integrating over limits, the constant, c, drops out. Please notice how remarkably close our guesstimate is to the finite true area of 2.

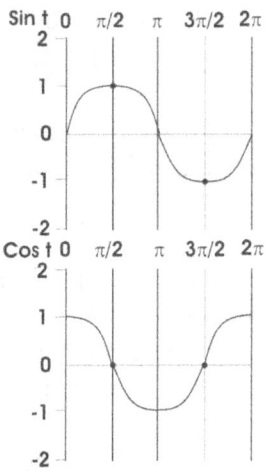

Notice that although we substitute the limits into the antiderivative expressions, which contain cosine expressions, we obtain the area under the <u>sine</u> curve. This is true because we always get the area under the function that is being integrated. Also, notice that we substituted an irrational limit, π, which we may expect to provide an inexact answer. But no, we get the exact answer, the integer, 2.

But most answers will be approximations due to engineering requests or the limitations of calculators and printers or the very nature of the number system.

Now, you are asked to find the area under the curve, $s' = 3t$, but above the t-axis and over the limits [0,3]. Here, you are given a curve that can be integrated. That is rather obvious because the prime symbol reminds us that we are working with a first derivative. These curves always integrate. It is now time for you to carry out some busy work yourself. But, you will be provided with the correct answer. You should arrive at an answer of 27/2 square units.

In the calculation of the sine curve area above, we restricted the limits to avoid dealing with negative areas below the horizontal axis.

Other Applications of Maxima and Minima
Other notations: f'(t) and f''(t) refer to 1st and 2nd derivatives

1. What is the maximum rectangular area that can be enclosed by a 25 foot long rope?
   The perimeter is given so create a function $25 = 2x + 2y$
   Simplify $12.5 = x + y$
   Establish the formula for the area $Area = xy$
   Reduce this expression to one independent variable $A = x (12.5 - x)$
      by solving for y in terms of x. Create the Area function, $A = 12.5x - x^2$
   Take the first derivative $A' = 12.5 - 2x$
   Calculate the critical value (where tangent slope = 0) $0 = 12.5 - 2x$
                                                               $x = 6.25$
   Calculate the value of y $y = 12.5 - 6.25$
      Height and width are equal so the rectangle is a square $y = 6.25$
   Test just to the left and right of the critical number, 6.25 $A = xy$
                                                               $A = 39.6$ units^2
   Employ the First Derivative Test, test with a 6, $12.5 - 2(6)$, gives a pos. slope
                  Test with 7, $12.5 - 2(7)$, gives a neg. slope
                  A pos. slope followed by a neg. slope implies a <u>max.</u>
   As a check, use the Second Derivative Test. We have $A'' = -2$ which is a neg. result.
   This indicates a decreasing rate of change for the slope which implies a <u>max.</u>

2. If two real numbers are related such that one is one less than twice the other, what is
their minimal product? Create a functional relationship from the given problem:
                                                               $P = x(2x-1)$
                                                               $P = 2x^2 - x$
   Get the first derivative $P' = 4x - 1$
   Set the slope to zero and solve for the critical number $0 = 4x - 1$ slope set
   The critical number is x=1/4, Employ test values 0 and 1,
   Employ the First Derivative Test, test with 0, $4(0) - 1$, gives a neg. slope
                  Test with 1, $4(1) - 1$, gives a pos. slope
         A neg. slope at the left followed by a pos. slope at the right implies a <u>min</u>
   Check by use of the 2nd Derivative Test. $P'' = 4$. A positive value implies a slope
that is increasing, +, so we have a <u>min.</u>
         Calculate the product of these two real numbers: $1/4 [(2)(1/4) - 4/4]$
            The two real numbers are ¾ and -3/2 $1/4 [-2/4]$
                                                               $- 2/16$ or
   The product of these two real numbers is: $- 1/8$
So we see that there is virtually no end to the problems that can be created and solved by
employing the first derivative, the second derivative and extrema. Many of these
problems are quite practical and calculus is usually the simplest approach in order to
produce a solution.

Trying Two Methods

We get further indication that the area under the curve is the area between the curve and the horizontal axis but bounded by the upper and lower limits of integration. We compare the results from both methods of calculation, the algebraic method and the calculus method employing the equation: $y = x + 2$

by Algebra:
Given: $A = \frac{1}{2}$ bh
$A = \frac{1}{2} (4 * 4)$
$A = 8$ units^2

by Calculus:
Area $= \int^2_{-2} (x + 2)\, dx$
$= x^2/2 + 2x\,]^2_{-2}$
$= 2 + 4 - \lceil (-2)^2/2 + 2(-2) \rceil$
$= 6 - (2 - 4)$
$= 8$ units^2

Now, don't complain about the extra length of the calculus calculation because the calculus method is perfectly general. It works even when the hypotenuse is curved by an equation such as: $y = x^2 + 2$

Area $= \frac{1}{2}$ bh
won't work
with this problem

Area $= \int^2_{-2} y\, dx$
Area $= \int^2_{-2} (x^2 + 2)\, dx$
$= x^3/3 + 2x\,]^2_{-2}$
$= 8/3 + 4 - (-8/3 - 4)$
$= 20/3 + 20/3$
$= 40/3$ units^2

Expressed as a decimal fraction, this rational number is non-terminating. Compare the answer to the graph. Does the answer make sense? It certainly does and gives us additional confidence in the calculus method. The graph is a parabola that is elevated two units above the origin. The preferred answer is in rational number form.

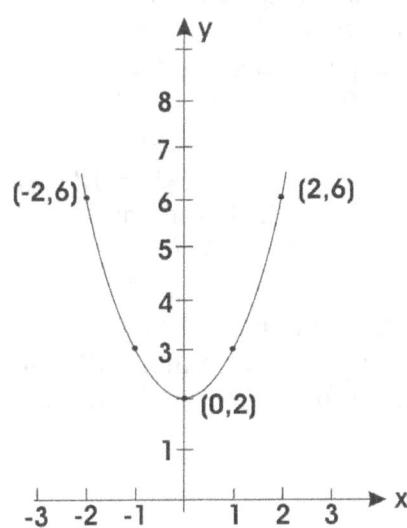

Consider the trapezoid over the range, $0 \leq x \leq 2$. The area calculated by Calculus is slightly less than the total area under the the twin trapezoids.

$1/2\ A = a\ (6 + 2)/2$
$\frac{1}{2}\ A = 2(\ 8\ )/2$
$A = 16$
$A = 48/3$

This is in accordance with our expectation.

Related Rates

The position equation gives the location of any point on the curve. The first derivative gives the rate of change of position and is called the velocity equation. The second derivative gives the rate of change of velocity and is called the acceleration equation. We can conclude that Calculus is much concerned about rates of change.

Sometimes, we have an equation which is concerned with several rate changes. Let us begin with the Pythagorean Theorem which deals with the relationship of the sides of a right triangle. In words, this equation is expressed as the square on the hypotenuse of a right triangle is equal to the sum of the squares of the other two sides. In symbols, $c^2 = a^2 + b^2$. Each letter represents the length of a side of the right triangle. If we consider these lengths to be variables, then we can differentiate each with respect to time and produce the differential equation that follows from the given equation.

$$c^2 = a^2 + b^2$$

Taking the first derivative with respect to time gives: $2c \, dc/dt = 2a \, da/dt + 2b \, db/dt$
Now, let us consider a problem in which this equation could be used. This is of the sliding ladder type. Of course, this could be a sliding glass panel or a sliding beam.
Suppose that a carpenter must mount a mirror to the wall and temporarily rests this mirror against the wall such that the following right triangle is formed. The mirror is 10 feet long and it rests in such a manner that the projected height along the wall is 8 feet and the length projected along the floor is 6 feet. If, due to gravity, the mirror starts to slide downward against the wall at a rate of -2 feet per second, at what rate does the bottom of the mirror move away from the wall?

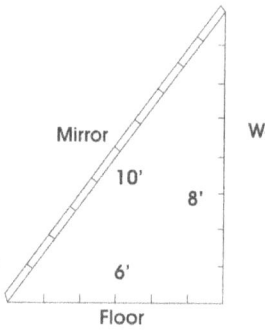

We are ready to plug all known values into our differential equation. Since c is the length of the mirror, it is a constant and the derivative of a constant is equal to zero. Substitute zero for dc/dt. The projected altitude is decreasing at the rate of -2 feet per second and the rate at which the projected base is increasing is the unknown value. $0 = 2a \, da/dt + 2b \, db/dt$

Make the substitutions $0 = 2(8)(-2) + 2(6) \, db/dt$
Solve for db/dt $0 = -32 + 12 \, db/dt$
$$32/12 = db/dt$$
$$2 \; 2/3 \; ft/sec = db/dt$$

Now, don't let the unknown, db/dt, psyche you out. Even though it is the ratio of differentials, it is simply the unknown rate that you are seeking. After differentiating the Pythagorean equation with respect to time, we simply plug in the known values and solve for db/dt (the unknown rate).

Applications of Maxima and Minima

Other notations: f′(t) and f″(t) refers to 1st and 2nd derivatives (the primes)

1. What is the maximum rectangular area that can be enclosed by a 16 foot long rope?
   Since the perimeter is given, create the function $16 = 2x + 2y$
   Simplify $8 = x + y$
   Establish the formula for the area $Area = xy$
   Reduce this expression to one independent variable $A = x(8 - x)$
      by solving for y in terms of x
   Create the Area function which is to be maximized $A = 8x - x^2$
   Get the first derivative $A' = 8 - 2x$
   Calculate the critical number (where tangent slope = 0 $0 = 8 - 2x$
      or undefined) $x = 4$
   Calculate the value of y $y = 8 - x$
   Height and width are equal so the rectangle is a square $y = 4$
                                 $A = 8x - x^2$ from above
      Test just to the left and right of the root, $x = 4$ $A' = 8 - 2x$
   Employ the First Derivative Test. First test with x = 3, 8 - 2(3), gives a positive slope.
   Next, test with x = 5, 8 - 2(5). This gives a neg. slope. These two results imply a <u>max.</u>
      As a check, use the Second Derivative Test $A'' = -2$
      The indication that the slope rate of change is decreasing, (-),implies a <u>max.</u>

2. If two real numbers are related such that
   one is three less than twice the other,
   what is their minimal product?

x	P
1	-1
2	2
3	9
0	0
-1	5

   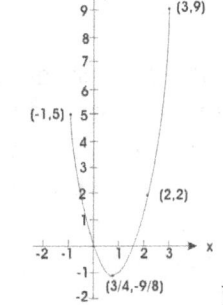

   Create a functional relationship from the given $P = x(2x - 3)$
      data $P = 2x^2 - 3x$
   Get the first derivative $P' = 4x - 3$
   Set the slope equal to zero and solve for the critical number $0 = 4x - 3$
   3/4 is a critical number, use test values just to the left and right $x = 3/4$
   Employing the First Derivative Test, first use ½,   4(1/2) - 3, gives a neg. slope
      next use 1,   4(1) - 3, gives a pos. slope, these two results imply a <u>min</u>
      Employ 2nd Derivative Test as a check, P″= 4,   The slope rate of change
                           is increasing, (+),   implies a <u>min.</u>
   The product of these two real numbers is: ¾ [(2)(3/4) - 12/4]
      The two real numbers are ¾ and -3/2   ¾ [-6/4]
      The product of these two numbers is: - 9/8

26

Position, Velocity and Acceleration

We should understand that any of the quantities in the caption above can be positive or negative. If a point lies to the left of zero on the horizontal axis, then it has a negative value for position. To the right of zero indicates a positive position. The velocity is negative if the velocity is directed towards the left on the horizontal axis or downward on the vertical axis. To the right on the horizontal axis or upwards on the vertical axis implies a positive velocity.

Acceleration is a bit more complex. The acceleration is positive if the object is moving either to the right or upward and speeding up or is moving in a negative direction and slowing down. Acceleration is negative if the object is moving in a positive direction and slowing down or is moving in a negative direction and speeding up.

Furthermore, it must be clearly understood that velocity is a vector quantity while speed is a scalar quantity. In other words, velocity has both magnitude and direction. Speed has magnitude only. We could say that speed is the absolute value of velocity.

If velocity and acceleration have the same sign, the object is speeding up. If velocity and acceleration have opposite signs, the object is slowing down.

It is time for us to explain why the 2nd Derivative Test producing a minus indicates a peak or maximum. On the other hand, when the 2nd Derivative Test produces a plus, a minimum (valley) is indicated.

Remember that the 2nd Derivative Test checks the rate of change of the velocity function. At the right, we simply graph the function and ask that you identify the peak and the valley. Draw some tiny arrows over the graph just to the left and right of the peak. Do the same over the valley and you will see an increasing slope as we approach from the left. This will trace out a valley. If the left approach traces out a peak, we have a decreasing slope indicating a maximum.

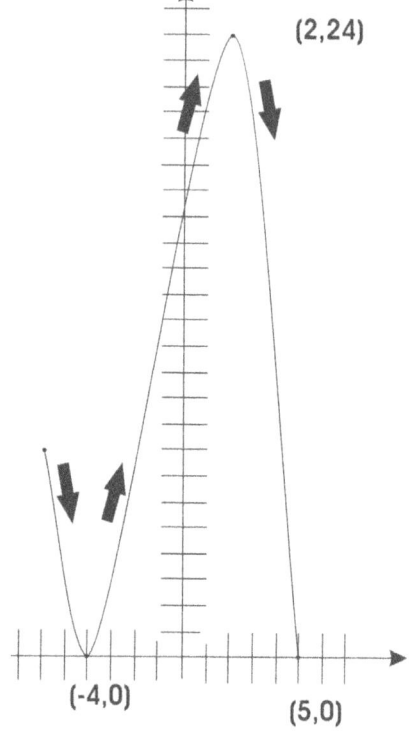

Important Differentiation Formulas

If we are seriously concerned about using Calculus instead of simply satisfying an intellectual curiosity, then memorization of the following is a must whether you have a photographic memory or not. When taking tests, you need instant recall of all the following. It is recommended that, instead of simply trying to memorize the symbols, commitment must be made to memorize the statements in words. This will enhance instant recall.

Let us agree upon the meanings of the following symbols. We will use f and g to represent functions instead of f(x) and g(x). The letters, m, n, b and k represent constants. The symbols, D_x, identify the differentiation operator.

Statement	Operator	Function	Derivative
The derivative of the sum of functions is the sum of the derivatives	D_x	$(f + g)$	$f' + g'$
The derivative of the difference of functions is the difference of the derivatives	D_x	$(f - g)$	$f' - g'$
The derivative of the product of a constant and a function is the product of the constant and the derivative	D_x	(kf)	kf'
The derivative of the product of a constant and the first power of the variable is simply the constant	D_x	(kx)	k
The derivative of the product of two functions is the first times the derivative of the second plus the second times the derivative of the first	D_x	(fg)	$fg' + gf'$
The derivative of the quotient of two functions is the denominator times the derivative of the numerator minus the numerator times the derivative of the denominator, all divided by the square of the denominator	D_x	(f/g)	$(gf' - fg')/g^2$
The derivative of the product of a constant and a power of the variable is the product of the power, the constant and the variable with the power reduced by one	D_x	(kx^n)	nkx^{n-1}
The derivative of a constant is equal to zero	D_x	(k)	0
The derivative of e^x is the same function (unique)	D_x	(e^x)	e^x
The derivative of a base to the variable power is the natural log of the base times the base to the variable power	D_x	(b^x)	$\ln b(b^x)$
The derivative of the natural log of x is the reciprocal of x	D_x	$\ln x$	$1/x$
The derivative of a base to a functional power is the natural log of the base times the derivative of the function times the base to the functional power	D_x	b^f	$(\ln b)f'b^f$

Differentiation Exercises

Remember, we are using f and g to represent functions instead of f(x) and g(x). m, n, b and k represent constants and the symbols, D_x, identify the differentiation operator. The symbol, ', identifies the first derivative.

Function	Reminder (if needed)	Derivative
$3x^4 - 4x^3 + 2x^2 + 5$	$\underline{D_x} (f + g) \rightarrow f' + g'$	$12x^3 - 12x^2 + 4x$
$2 - 5x + \sqrt{x}$	same as above	$-5 + \frac{1}{2}(x^{-1/2})$
$\ln x$	D_x of natural log of x gives reciprocal of the variable	$1/x$
$\ln(3x + 1)$	f'/f	$3/(3x + 1)$
$\ln(\tan x)$	f'/f	$\sec^2 x/\tan x$
$\ln(4x^2 + 3x - 2)$	f'/f	$(8x + 3)/(4x^2 + 3x - 2)$
e^{4x-3}	$f' e^f$	$4e^{4x-3}$
$\log_5 x$	$1/(x \ln b)$	$1/(x \ln 5)$
2^x	$b^x (\ln b)$	$2^x \ln 2$
$\sin x$		$\cos x$
$\cos x$		$-\sin x$
$\tan x$		$\sec^2 x$
$\sec x$		$\sec x \tan x$
$\cot x$		$-\csc^2 x$
$e^k - 4t^3$	e^k is a constant	$-12t^2$
$\cos t - 4/t^3$	$D_t(f + g) \rightarrow f' + g'$	$-\sin t - 12/t^4$
$x^3 - 2x^2 - x + 3$	$D_x(f + g) \rightarrow f' + g'$	$3x^2 - 4x - 1$
$\ln 6 + 4x^{-5}$	ln 6 is a constant	$-20x^{-6}$

Solutions to Differentiation Problems
D_x, dy/dx, y′   All indicate the taking of the first derivative of the given function

Function

$(4x-3)^{-2}$ Find the first derivative   Employ the Power Rule
$$-2(4x-3)^{-3} \, D_x(4x-3) = -2(4x-3)^{-3} \, (4) = -8(4x-3)^{-3}$$

$(\sin x)/x$ Find the first derivative   Employ the Quotient Rule
$$[x \, D_x(\sin x) - (\sin x) \, D_x(x)] / x^2 = [x \cos x - \sin x \, (1)] / x^2 = [x \cos x - \sin x] / x^2$$

$x \sin x$ Find the first derivative   Employ the product Rule
$$x \, D_x (\sin x) \ + \ \sin x \ D_x \, x = \ x \cos x \ + \sin x \ (1) \ = \ x \cos x \ + \ \sin x$$

$\sin^3 (3x + 4)$ Find the first derivative   Employ the Power Rule
$$3 \sin^2 (3x + 4) \, D_x(3x + 4) = \ (3) \, 3 \sin^2 (3x+4) = 9 \sin^2 (3x+4)$$

$y = (x + 3)^{1/2}$ Find the first, second, third and fourth derivatives (all by Power Rule)
$$y' = \tfrac{1}{2} (x + 3)^{-1/2} \, D_x(x+ 3) = \ \tfrac{1}{2} (x + 3)^{-1/2} (1) = 1/2(x + 3)^{-1/2}$$
$$y'' = -1/4(x+3)^{-3/2} \, D_x(x+3) = -1/4(x+3)^{-3/2} \, (1) = -1/4 \, (x+3)^{-3/2}$$
$$y''' = 3/8(x+3)^{-5/2} \, (1) = 3/8(x+3)^{-5/2}$$
$$y'''' = -15/16 \, (x+3)^{-7/2} \, (1) = 15/16 \, (x+3)^{-7/2}$$

$y = x(x-1)^3$ Find the critical numbers   Employ the Product Rule
$$xD_x(x-1)^3+(x-1)^3 \, D_xx = (x)3(x-1)^2(1)+(x-1)^3(1) = 3x(x-1)^2 + (x-1)^3 = (x-1)^2[4x-1]$$
Set slope = 0, then x = 1 or x =1/4   f′(x) is positive on both sides of x=1 indicating an inflection point at x=1 (point at which the concavity changes).

$y = 4x^3 - 8x^2 + 1$ Find the critical numbers   Employ the Power Rule
$$y' = 12x^2 -16x \ = 4x(3x-4)$$ Set y′ equal to zero and x = 0 or x = 4/3

$y = \sin x - \cos x$ on $[0, \pi]$ Find the solution in the given range. Employ the Power Rule.
$$y' = \cos x + \sin x$$ Set y′ equal to zero and   sin x = - cos x   The only solution
in the given range is $3\pi/4$ See Appendix B

$y = x^4 - 18x^2 + 9$ Find the critical numbers
 $y' = 4x^3 -36x$ Set y′ equal to zero and factor
 $0 = 4x(x-3)(x + 3)$ The critical numbers are x = 0, x = 3, and x = -3
 $y'' = 12x^2 -36$
 $y'' (0) = -36$ indicates a max
 $y'' (3) = 72$ indicates a min
 and $y'' (-3)= 72$ indicates a min

Basic Integration Formulas

(Always <u>add</u> <u>a</u> <u>general</u> <u>constant</u> unless the true constant is known)

$\int [f(x) + g(x)] \, dx = \int f(x) \, dx + \int g(x) \, dx + C$
    The integral of a sum of terms is the sum of the integrals followed by an added constant

$\int k \, f(x) dx = k \int f(x) dx + C$
    The integral of the product of a constant and a function is the product of the
        constant and the integral of the function followed by an added constant

$\int ax^n \, dx = a \int x^n \, dx = a \, x^{n+1}/(n+1) + C$
    The integral of the product of a constant and a variable raised to a power is the
        constant times the variable raised one power divided by the new power and
        followed by an added constant

$\int (a/x) \, dx = a \int dx/x = a \ln |x| + C$
    The integral of a constant times the reciprocal of x is the product of the constant and
        the natural log of the absolute value of x and followed by a constant
        (only positive numbers have logs)

$\int e^{ax+b} dx = e^{ax+b}/a + C$ for the constants a and b with $a \neq 0$, the integral is the
exponential divided by the coefficient of the variable and followed by an added constant

$\int -\sin x \, dx = \cos x + C$

$\int \sec^2 x \, dx = \tan x + C$

$\int (\sec x \tan x) \, dx = \sec x + C$

$\int (-\csc x \cot x) dx = \csc x + C$

$\int -\csc^2 x \, dx = \cot x + C$

The above trigonometric integrals are the reverse of the differentiation formulas introduced
earlier. Some of the integrals of the inverse trig functions follow:

$\int dx/(\sqrt{1-x^2}) = \arcsin x + C$

$\int -dx / \sqrt{1-x^2} = \arccos x + C$

$\int dx /(x^2 + 1) = \arctan x + C$

$\int dx/(x^2 +1) = -\operatorname{arccot} x + C$

Integration Exercises

Indefinite Integrals (no limits)   **Answers**

$\int 4dx/3x^4 = (4/3)x^{-3}/-3 + C$ $-4/9\ x^{-3} + C$

$\int dx\ /\ x$ $\ln |x| + C$

$\int 4x^3\ dx\ /9 = -4/9\ x^4\ /\ 4 + C$ $(-1/9)\ x^4 + C$

$\int 5x^{1/3}\ dx =$ $15/4x^{4/3} + C$

$\int e^{5x+\pi}dx =$ $e^{5x+\pi}/5 + C$

$\int (-2/3)\cos t + 4/t)\ dt =$ $(-2/3)\sin t + 4 \ln |t| + C$

$\int -5 \sec^2 x\ dx =$ $-5 \tan x + C$

$\int -6x^{11}\ dx =$ $-1/2\ x^{12} + C$

$\int (\sec^2 x + 5/3\ x^{-1})dx =$ $\tan x + 5/3\ \ln |x| + C$

$\int -3\ /\ (1 - x^2)^{1/2}\ dx =$ $-3\ \arcos x + C$

Definite Integrals (limits employed)

$\int^3_{-1}(4x^2 -3x + 2)dx = 4\ x^3/3 - 3x^2/2 + 2x\ ]^3_{-1} = 36 - 27/2 + 6 - [\ 4/3 - 3/2\ -2] =$ $100/3$

$\int^{\pi/4}_0 \cos x\ dx = \sin x\ ]^{\pi/4}_0 = \sqrt{2}/2 - 0 =$ $\sqrt{2}\ /\ 2$

$\int^{16}_0 x^{3/2}\ dx = 2/5\ x^{5/2}\ ]^{16}_0 = 2/5\ [(16)^{5/2} - 0] = 2/5\ (4^5) = 2/5[(16^2)(4)] =$ $2048/5$

$\int^{\pi/3}_0 \sec^2 x\ dx = \tan x]^{\pi/3}_0 = \tan \pi/3 - \tan 0 = \sqrt{3}/1 - 0 =$ $\sqrt{3}$

$\int^2_0 (3x^2 + 2x)\ dx = 3x^3/3 + 2x^2/2\ ]^2_0 = 8 + 4 - [0] =$ 12

$\int^4_0 e^{-2x}dx = -1/2\ e^{-2x}\ ]^4_0 = -1/2\ e^{-8} - [\ -\frac{1}{2}\ e^0] \approx -0 + \frac{1}{2}(1)$ $\approx 1/2$

Integration Techniques

At this point, the only integration techniques that we have been exposed to are the Power Rule and the functions whose integrals have been memorized such as e^x and cos x. We need exposure to the u-substitution method to broaden our base of methods. In this method, we look for a piece of the function whose derivative also appears in the function. This is possibly in the denominator or something that is raised to a power. Now, this is a trial and error method so if at first you don't succeed try and try again. Set the symbol, u, equal to this selection, u = 5x, and take the derivative. As examples, consider:

$\int \cos 5x \, dx$

Out of desperation and frustration, let us try: u = 5x,

then du = 5 dx and du/5 = dx

Make substitutions $\int \cos u \, (du/5)$

$1/5 \int \cos u \, du = \quad 1/5 \sin u + C \quad = 1/5 \sin 5x + C$

$\int dx/(7x-3)$ let u = 7x-3, then du/dx = 7 and du/7 = dx

Make substitutions $1/7 \int du/u = 1/7 \ln |u| + C = 1/7 \ln |7x-3| + C$

$\int (x^3 \, dx)/(x^4-3)$ let u = x^4-3, then du/dx = $4x^3$ and du = $4x^3$ dx

Make substitutions $\int (x^4-3)^{-1} (4x^3) dx = \int du/u = \ln |x^4-3| + C$

$\int \cot x \, dx = \int (\cos x / \sin x) \, dx$ let u = sin x, then du/dx = cos x and du = cos x dx

Make substitutions $\int du/u = \ln |u| + C = \ln |\sin x| + C$

Note: only positive numbers have logs

Fundamental Integration Formulas (Where u is a function of x)

The Power Rule $\int u^n \, du = u^{n+1}/(n+1) + C$ were n ≠ -1

$\int du/u = \ln |u| + C$

$\int e^x = e^x + C$

$\int e^u \, du = e^u + C$

Examples:

$\int_0^{\pi/4} \cos x \, dx = \sin x \,]_0^{\pi/4} = \sin \pi/4 - \sin 0 = \sqrt{2}/2 - 0 = \sqrt{2}/2$

$\int x^2 \, dx/\sqrt{(x^3+4)}$ Then u = (x^3+4) du/dx = $3x^2$ and du = $3x^2$ dx

du/dx = $3x^2$ du=$3x^2$ dx

$1/3 \int du/\sqrt{u} = 1/3 \int (u^{-1/2}) du = 1/3 \, u^{1/2}/1/2 + C = 2/3(x^3+4)^{1/2} + C$

$\int e^x e^{2x} \, dx = \int e^{3x} \, dx$ Let u = 3x Then du = 3dx and $1/3 \int e^{3x} \, 3dx =$

$1/3 \int u \, du = 1/3 \, e^{3x} + C$

The First and Second Derivatives and Graphing

Being able to calculate the first derivative is a tremendous help in graphing. Remember that the first derivative gives the slope of the tangent line at any point on the curve. In other words, the first derivative points in the direction of the curve at any point on the curve. Let us work with a fourth degree equation:

$f(x) = x^4/4 - x^2/2 + 2$ $f'(x) = x^3 - x$ $f''(x) = 3x^2 - 1$

Set $f'(x) = 0$ and factor

$0 = x(x+1)(x-1)$

The critical numbers are: -1, 0, 1

Check: $f'(-1) = -1+1 = 0$ OK
         $f'(0) = 0$ OK
         $f'(1) = 1 - 1 = 0$ OK

$f(-1) = ¼ - ½ + 2 = 7/4$ Now, at the left, plug these critical numbers into f(x)
$f(0) = 2$ in order to produce three critical points
$f(1) = ¼ - ½ + 2 = 7/4$ on the graph of the function, f(x).

This gives us the three critical points to place

The coordinates of the   on our graph of this original function.
three critical points   Next, include on this graph
are: (-1, 7/4), (0, 2)   the three corresponding critical numbers.
and (1, 7/4).   Place the three critical numbers below the horizontal axis. Include the two test values, -2 and 2.

The second derivative tests, at the right, display positive $f''(-1) = 3-1$
values at -1 and 1 indicating minima. The negative value $f''(0) = -1$
at $x = 0$ indicates a maximum. Visually, check the graph. $f''(1) = 3-1$

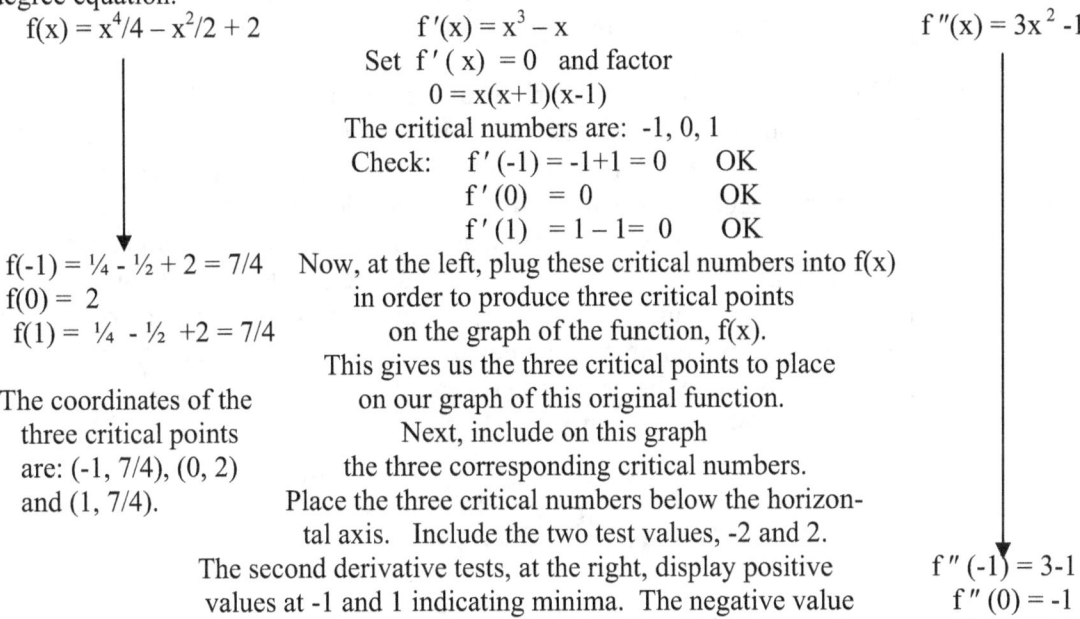

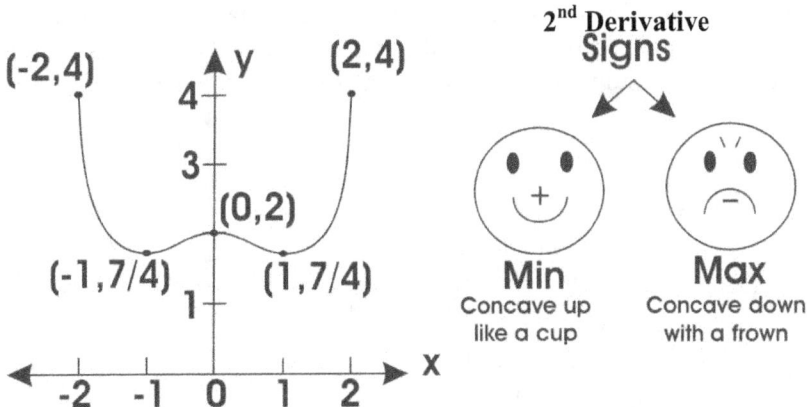

Next, we need to choose test values to the left and right of each critical number. To simplify, let us choose -2, -1/2, 1/2, and 2. Now, let us find the slope directions associated with these four test x-values because we want to employ the First Derivative Test as an alternative test. In order to accomplish this, we must plug these four test x-values into the first derivative of our function. This gives the nature of the slopes to the left and right of each critical number? This detects the extrema, if any.

The First Derivative & Graphing

$f'(x) = x^3 - x$ The first derivative equation

f '(-2) = -8 +2   - over the half open interval, $-2 \leq x < -1$, implies a negative slope

f '(-1/2) = -1/8 + ½   + over the interval, $-1 < x < 0$,   implies a positive slope

f '(1/2) = 1/8 - ½   - over the interval, $0 < x < 1$,   implies a negative slope

f '(2) = 8 -2   + over the half open interval $1 < x \leq 2$, implies a positive slope

Once again we are displaying the graph to clearly show the relationships. The 2nd Derivative Test is simpler but sometimes it fails by producing a zero. In this case, we must go back to the 1st Derivative Test which means that we must choose test values to indicate the tangent directions over selected intervals. Notice that we do not include closed intervals at the critical numbers because we know that those slopes are zeros and are the separators of the positive and negative slopes. Obviously, in regard to the critical numbers, a negative slope before the critical number and a following positive slope points to a minimum. The positive slope preceding and the negative slope following points to a maximum. Once again, we provide the graph so you can visually check.

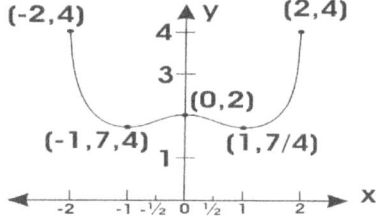

Since we have been illustrating primarily with continuous and differentiable functions, we now need to realize that some functions are not continuous at certain x-values and some functions, although continuous at all x-values are not differentiable at some x-values. Perhaps we can best indicate this fact by displaying some of these curves for which you should be on the lookout..

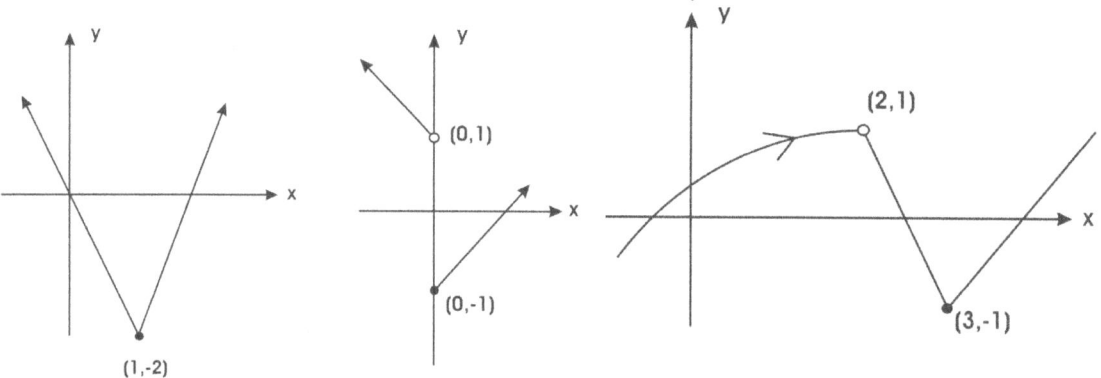

Continuous but not   Jump discontinuity at   Not continuous at x=2
differentiable at x = 1.   x = 0   Not differentiable at x=3

Left-hand and right-hand limits, Continuity

Polynomial functions are easy to work with. That is why our original concentration was on these functions. They are everywhere continuous which means that there are no missing points on these graphs. We must introduce you to some special terminology so you can deal with the functions that have gaps or missing points on their graphs Consider the function, $f(x) = x - 1$. This curve has been produced by not lifting the pen from the paper. As a result, there are no missing points on the curve and we say that this is a continuous function. We must prepare you for the exceptional cases that may occur. Perhaps, the best way to introduce these exceptions is by the use of graphs.

We can approach a point on a graph from the left or from the right. Let us consider the four graphs on the next page. We will employ a special notation to indicate that we are approaching a point on the graph from the left.   Also, a special notation will indicate an approach from the right.   Besides, we need to carefully distinguish between the x-value and the functional value (height) of the graph. We may refer to these two values as the x-value and the functional value.   Given the function, $f(x) = x-1$, and an x-value of 0.3, then the functional value is equal to -0.7. The large graph on the next page is a concise representation of the interplay between the x-values and the corresponding functional values. You are asked to intensely study all the graphs on the next page,

the function	from the left	from the right
$f(x) = x - 1$	$f(x) = -0.7$	$f(x) = -0.7$
	$\lim x \to .3^-$	$\lim x \to .3^+$

The symbol, -, employed as a superscript indicates an approach from the left. An approach from the right is indicated by employing the symbol, +, as a superscript. In each case, the limit is the value (height) reached by the graph of the function as x approaches the value, .3. If these two one-sided limits are equal, then we say that the general limit exists in the neighborhood of x = .3. This is the case in the given example. If the two limits are unequal, then the limit does not exist in the neighborhood of x = 0.3.   But equality of the one-sided limits is not sufficient to prove that the function is continuous at this x-value under consideration.   Additionally, we must show that the function is defined at that precise x-value and the functional value (the ordinate) is equal to the general limit. This guarantees continuity at the specified x-value.

The four graphs on the next page should be referenced as this page is read. The two pages supplement each other and their importance should not be overlooked. For emphasis, we will enlarge only eleven of the infinite number of points on the large graph and associate each with a coordinate pair, such as (.3,-.7). Actually, any point on this graph is associated with a coordinate pair but we have enlarged only eleven of these.

If the general limit does not exist at a specific x-value, then there is a discontinuity at this x- value. Also, at any break in the graph, the function is not differentiable. If the graph has an abrupt change in direction at a corner (a cusp), the general limit does not exist and the graph would be discontinuous at that x-value.

Graphs, Limits and Continuity

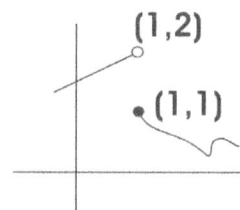

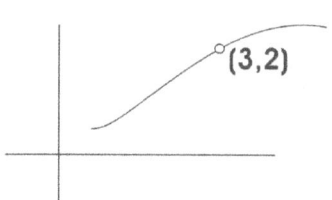

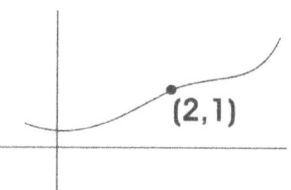

At x = 1
    left-hand limit exists
    right-hand limit exists
    one sided limits are not
      equal
    general limit does not
      exist at x = 1
    not continuous at x = 1

At x = 3
    left-hand limit exists
    right-hand limit exists
    one-sided limits are
      equal
    general limit exists
    not defined at x = 3
    not continuous at x = 3

At x = 2
    left-hand limit exists
    right-hand limit exists
    one sided limits are
      equal
    general limit exists
    defined at x = 2
    continuous at x =2

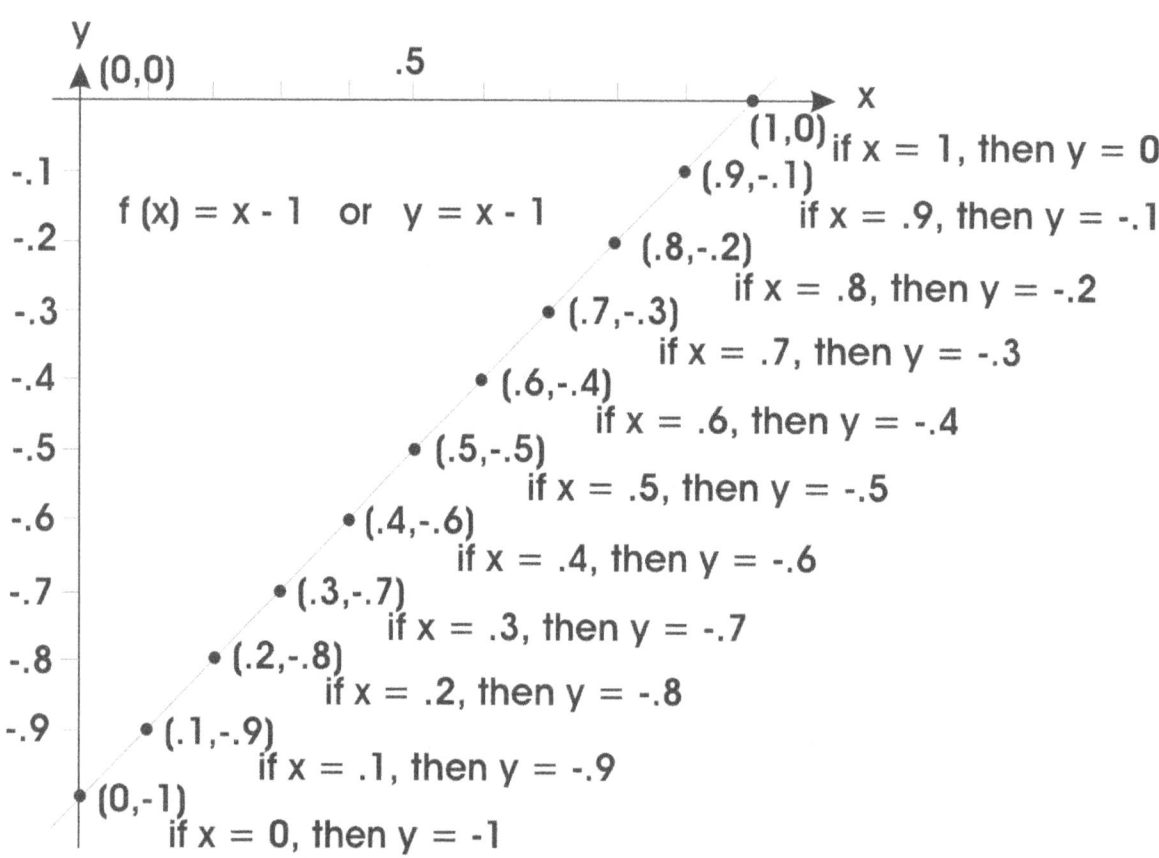

Theorems

At x = 2
  The function is defined at x = 2
  The limit exists in the neighborhood
  The limit is equal to the function at x = 2
  The curve is continuous at x = 2

At x = 3
  Even though defined at x = 3,
    the limit does not exist (cusp)
  The function is discontinuous at x = 3
    and not differentiable at x = 3

At x = 5
  The limit exists but the curve is not
    defined at x = 5
  The curve is neither differentiable or
    continuous at x = 5.

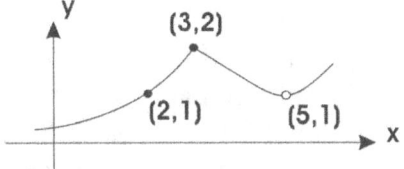

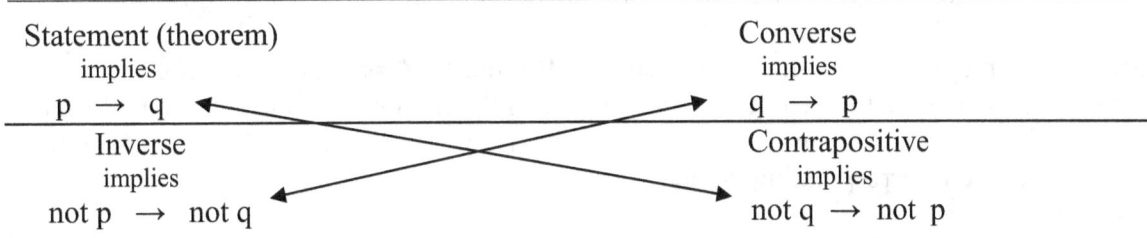

Statement (theorem) implies $p \to q$	Converse implies $q \to p$
Inverse implies not p $\to$ not q	Contrapositive implies not q $\to$ not p

At a point where the function is continuous but not differentiable, there will be an abrupt change of direction. This is illustrated at x = 3 above. Also, the function is not differentiable wherever there is a break in the graph. The point (5,1) does not exist The graph is not defined at x = 5.

The theorem and it's contrapositive must always agree.
The converse and the inverse must always agree
  Theorem: If a function is not continuous at x = c, then it is not differentiable at x = c.   T
  Converse: Not differentiable at x = c, then it is not continuous at x = c.   F
  Inverse: If a function is continuous at x=c, then it is differentiable at x = c.   F
  Contrapositive: Differentiability at x = c implies continuity at x = c.   T
Intuitively, a continuous function is graphed by not lifting your pen from the paper. But additionally, there should be no sharp corners (cusps). The mathematical definition implies the following:
1. The function is defined at any x-value, x = c. (At any x-value, there exists but one ordinate)
2. The right-hand and left-hand limits must equal the ordinate (height) at x = c... Many functions are continuous at any x = c. These include polynomials, exponential, logarithmic and rational functions.

Some functions are removably discontinuous at x = c if the limit exists in the neighborhood of x = c. In other words, a point discontinuity exists. We can define the missing point and establish continuity at that x-value. This cannot be done at x-values where jump or infinite ordinates occur.

D'Alembert Provides the Answer

Jean Le Rond D'Alembert

A quantity is something or nothing. If it is something, it has not yet vanished; if it is nothing, it has literally vanished. The supposition that there is an intermediate state between these two is a chimera.

Bishop Berkley was puzzled by the infinitesimals. Were the strange quantities real numbers or figments of the imagination? D'Alembert, the creator of the limit concept, answered Bishop Berkeley's question in no uncertain terms. It was the genius of D'Alembert that removed the stigma from calculus and allowed it to become a completely respected branch of mathematics. If the general limit did not exist, then this function was discontinuous for that certain x-value. If the general limit did exist, then at least the discontinuity was removable simply by definition.

Still, division by zero could not be carried out. But the expression, 0/0, was not illegal. It simply was said to be indeterminate. Working with the right side of the equation allowed us to explore the many ordinate values that may exist over the domain of the x-values as they generate the corresponding graph heights (the ordinates).

In other words, the differential is a tiny nonzero number unless we are in the process of taking the limit. This means that we can multiply or divide by differentials just as we do with any nonzero real number. It is only during the limiting process that we must avoid dividing by zero when the numerator is a nonzero real number.

The limit concept allows us to dispose of bishop Berkeley's "ghosts of departed quantities". At the same time, the good Bishop made an important contribution to the advancement of mathematics. He forced mathematicians to be more precise in their arguments. No longer could professional mathematicians deride the bishop. In time, D'Alembert's limit concept was recognized as the supreme test to end current arguments in calculus.

Two of the most productive mathematicians in calculus had unimpressive births. Isaac Newton was born prematurely on Christmas day in 1642. He came into this world so tiny that he was able to fit into a quart pot. This was only two months after his father, who tilled the soil, had died.

In 1717, a fondling was found on the steps of the church of Saint Jean Baptiste Le Rond in Paris. He was named Jean Le Rond and he later accepted the surname, d'Alembert. Much later, it was discovered that his birth mother was an aristocrat and his father was a general.

Obviously, how you come into this world is not as important as how you leave it. They were two of the greatest mathematicians.

In Conclusion

All polynomials are continuous everywhere.
All exponential functions are continuous everywhere.
Rational functions are continuous everywhere except at zeros of the denominator.
 All logarithmic functions are continuous over the entire domain $(0,\infty)$.
The six basic trigonometric functions are continuous over their domains (all real
numbers for the sine and cosine functions). Restrictions are on the tangent function.

Helpful topics are included in the Appendices. Essential elements in algebra, trigonometry
and analytic geometry are summarized in the rear of the text so they may be quickly
accessed when needed.

In any presentation, there must be sufficient repetition.   For most students, it appears that
the magic number is three. Hopefully, our booklet will refer to a topic on at least three
occasions and encourage pencil-pushing throughout.

Begining calculus certainly should include a variety of applications. This is normally a
weak point in most presentations. Normally, students do not appreciate the full
significance of prolonged pencil-pushing. On a test, a student must pencil-push throughout
so this must be a vital part of the learning process. Perhaps a part of this is in the arm.

Refer to Appendix B, the trigonometry refresher. Text books commonly employ a half
dozen pages to express less than one-half of this material. Can you believe that a similar
claim can be made for Appendix C, exponentials and logarithms. Can any textbook page
match the factual output of Appendix H, graphs of common functions?

Other examples of factual conciseness is expressed in Appendix A and Appendix I which
deal with literal equations. These Appendices provide the opportunity for the student to
duplicate both through pencil-pushing. Do this several times. This will be a confidence
booster.

The author hopes that this booklet will provide you with the basic skills needed to
successfully complete a course in calculus. To pursue Calculus beyond this booklet,
consider ordering several of the books listed in the bibliography on page ix. Also, perhaps
this exposure will encourage students to pursue other courses that employ differential and
integral calculus.

Those of you who pursued this course, due to intellectual curiosity, must be delighted to
realize that now you know how and why calculus was created. You must also realize that
this was truly one of the greatest accomplishments of the human intellect. Congratulations
and best wishes in your pursuit of understanding a truly great accomplishment in
mathematics and logic.

Appendix A

Literal Equations

Algebra – Solving Equations in Physics

Albert Einstein asked his uncle, who was an engineer, "What is algebra?" The answer was a classic. The uncle said, "Algebra is a kind of arithmetic, in which, if you don't know the value of a number, you call it "x" and pretend that you know it."

In physics, average speed is equal to the distance divided by the time, $v_a, = s / t$. Starting with this equation, solve for the distance, s:

Get the desired unknown on one side of the equation by itself:

Solve for s	$v_a = \ s / t$
Equals multiplied by equals gives equals	$t \, v_a = t \, (s / t)$
Any number divided by itself equals 1, except zero	$t \, v_a = 1 \, s$
Any number multiplied by 1 is unchanged, rotate 180,	$s = t \, v_a$
on both sides, maintain left to right sequence of symbols	

Get the desired unknown on one side of the equation by itself:

Solve for t	$v_a = \ s / t$
Equals multiplied by equals gives equals	$t \, v_a = t \, (s / t)$
Equals divided by equals gives equals	$t \, v_a / v_a = t \, (s / t) / v_a$
Any number divided by itself equals 1, except zero	$1 \, t = s / v_a$
Any number multiplied by 1 is unchanged	$t = s / v_a$

Use the following to get the desired unknown on one side of the equation by itself:

Equals	$6 = 6$	•	Equals	$6 = 6$
Equals multiplied by equals	$6 * 2 = 6 * 2$	•	Equals subtracted from equals	$6\text{-}2 = 6\text{-}2$
Gives equals	$12 = 12$	•	Gives equals	$4 = 4$

Equals	$6 = 6$	•	Equals	$6 = 6$
Equals divided by equals	$6 / 2 = 6 / 2$	•	Equals added to equals	$6 + 2 = 6 + 2$
Gives equals	$3 = 3$	•	Gives equals	$8 = 8$

Appendix B

Trigonometry Refresher

You must learn the ratios of the sides of five basic right triangles. Committing these to memory is quite easy to do because only four numbers are employed. Memorize the angular measures, the triangles, the function formulas and at least the two top graphs.

θ			Pythagorean Theorem	Functions	
R	deg				

$0 \qquad 0°$

$$c^2 = a^2 + b^2$$
$$c^2 = 0 + 4$$
$$c = 2$$

$$\sin \theta = y/c$$
$$= 0/2$$
$$= 0$$

$\pi/6 \qquad 30°$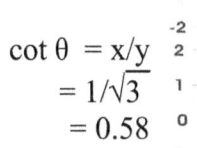

$$c^2 = a^2 + b^2$$
$$c^2 = 1 + 3$$
$$c = 2$$

$$\cos \theta = x/c$$
$$= \sqrt{3}/2$$
$$= .87$$

$\pi/4 \qquad 45°$

$$c^2 = a^2 + b^2$$
$$c^2 = 2 + 2$$
$$c = 2$$

$$\tan \theta = y/x$$
$$= \sqrt{2}/\sqrt{2}$$
$$= 1$$

$\pi/3 \qquad 60°$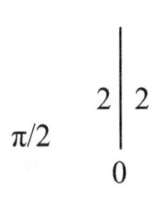

$$c^2 = a^2 + b^2$$
$$c^2 = 3 + 1$$
$$c = 2$$

$$\cot \theta = x/y$$
$$= 1/\sqrt{3}$$
$$= 0.58$$

$$\sec \theta = c/x$$
$$= 2/1$$
$$= 2$$

$\pi/2$

$$c^2 = a^2 + b^2$$
$$c^2 = 4 + 0$$
$$c = 2$$

$$\csc \theta = c/y$$
$$= 2/2$$
$$= 1$$

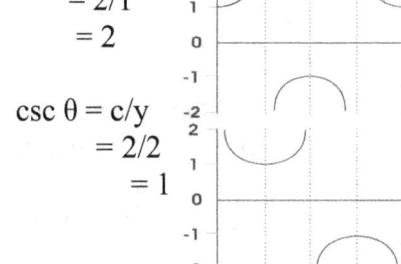

Appendix C

Logarithms Base 10 & Exponentials

Log of a product =   Sum of the logs
  $\log 4 = \log 2*2 = \log 2 + \log 2 = .301 + .301 = .602$
  $\log 6 = \log 2*3 = \log 2 + \log 3 = .301 + .477 = .778$
  $\log 8 = \log 2*2*2 = \log 2 + \log 2 + \log 2 = 0.903$
  $\log 10 = \log 2 * 5 = \log 2 + \log 5 = .301 + .699 = 1$
Log of an Exponential = the exponent * log of the base
  $\log 2^3 = 3 \log 2 = 3 * .301 = 0.903$
  $\log 9 = \log 3^2 = 2 \log 3 = 2 * .477 = 0.954$

Forms of Numbers

Standard	Exp	Log$_{10}$
100	10^2	2
10	10^1	1
1	10^0	0
.1	10^{-1}	-1
.01	10^{-2}	-2
.001	10^{-3}	-3

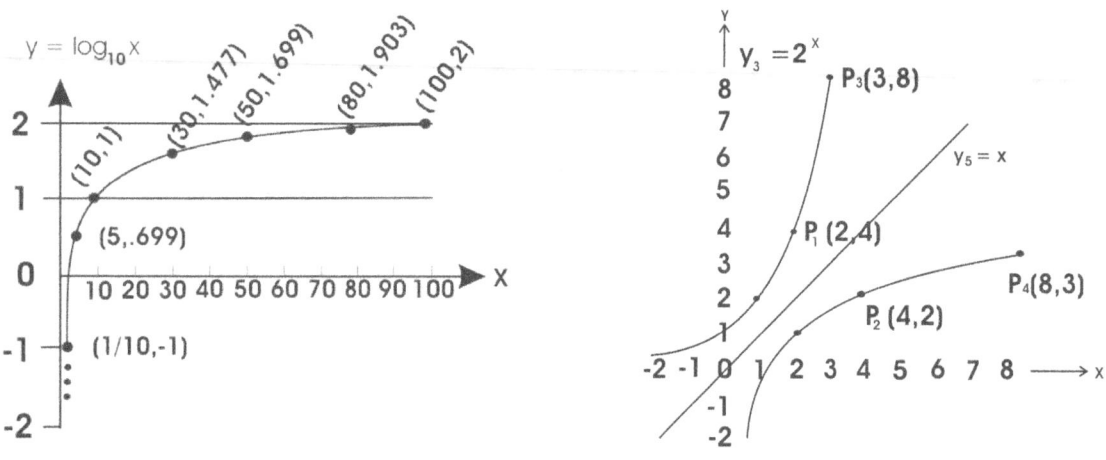

Graph of Common Logarithms (base 10)

Exponential & logarithmic (base 2)

logs of numbers > 0

numbers	log$_{10}$
1/10	-1
1	0
2	.301
3	.477
4	.602
5	.699
6	.778
7	.845
8	.903
9	.954
10	1

Exponentials of the same base
  The product of two exponentials = base$^{\text{sum of the exponents}}$
    $a^p * a^q = a^{p+q}$ $a^2 * a^5 = a^7$
  The quotient of two exponentials = base$^{\text{difference of exponents}}$
    $a^q / a^p = a^{q-p}$ $a^5 / a^2 = a^3$
  Base to a negative power = remove minus sign and place all below
    $a^{-q} = 1 / a^q$ $1 / a^5 = 1 / (a * a * a * a * a)$
  The exponential raised to a power = base raised to product of powers
    $(a^q)^p = a^{q*p}$ $a^{5*2} = a^{10}$
      Remember $a^2 a^4 = a^6$ and $(a^2)^4 = a^8$

Appendix D

Trigonometry
Functions and Inverse Functions

For each circular function, there exists an inverse function such that one is the reflection of the other in the line y = x which bisects the first and third quadrants. These are created by suitably restricting the domain of the function or the range of the corresponding inverse function. In any case, each axis is measured by the elements in the set of real numbers because we are now expressing angles in radians, not degrees. This means that both axes are graduated in real numbers with π about equal to 3.1416. With these specifications, we can graph a function and its inverse function on the same coordinate axes.

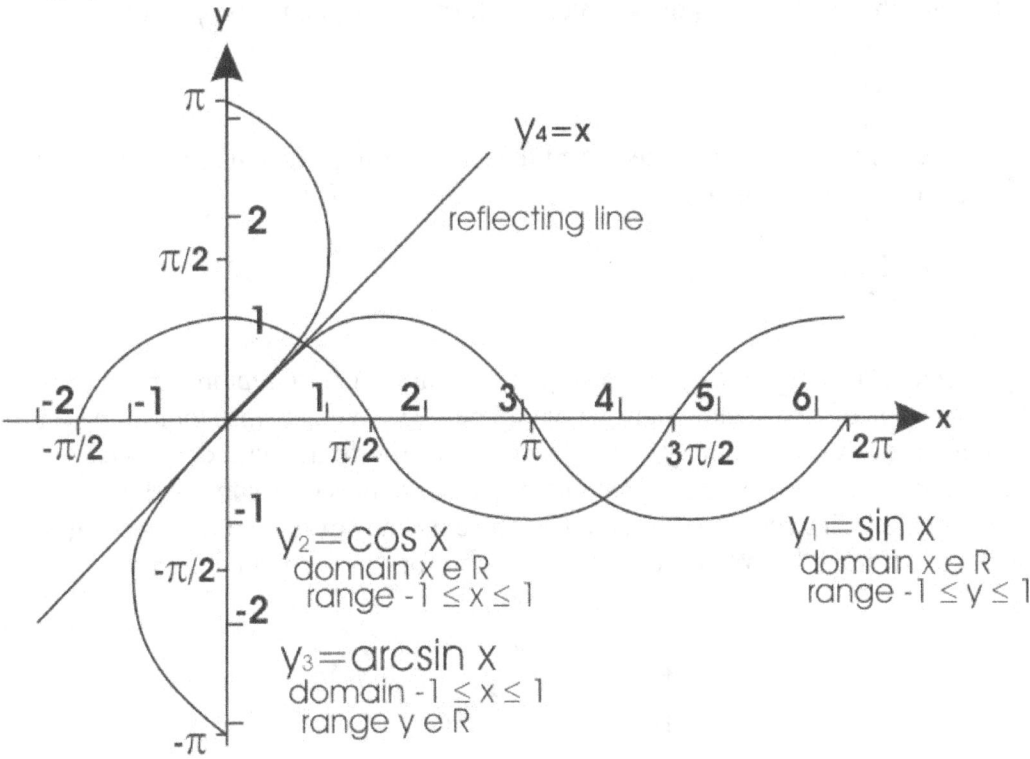

Appendix E

Essential Linear Elements
Given two points in the plane, $P_1(x_1,y_1)$ and $P_2(x_2,y_2)$, the slope of the line is given by:
$$m = (y_2 - y_1) / x_2 - x_1)$$

We may easily place this formula into a very useful form called the slope-y-intercept form. Given the two points, $P_2(5,7)$ and $P_1(3,1)$, the slope of this line, l_1, is called m_1. $m_1 = (7-1) / (5-3) = 3$. This is simply the ratio of rise to run which architects commonly use to specify roof slope.

Now that the slope of the line has been calculated, we may use this slope with either of the two points to place the equation in slope-y-intercept form. First, let us use point $P_1(3,1)$.

$$y - y_1 = 3 (x - x_1)$$
$$y - 1 = 3(x - 3)$$
$$y = 3x - 9 + 1$$
$$y = 3x - 8 \qquad \text{y is at the left, at the right the slope is 3 and y-intercept is -8}$$

Now, we should get the same result by using the point $P_2(5,7)$.

$$y - y_1 = 3(x - x_1)$$
$$y - 7 = 3 (x - 5)$$
$$y = 3x - 15 + 7$$
$$y = 3x - 8$$

A line, l_2, perpendicular to the line, l_1, has a slope that is the negative reciprocal of the slope of 3 so the slope of a perpendicular line would be -1/3. We use this slope along with one point on the line of the equation, $y = 3x - 8$. There are an infinite number of lines perpendicular to the equation, $y = 3x - 8$, but one point that comes easily to mind is the point $P_3(0,-8)$ since a y-intercept of -8 guarantees the existence of this point. Once again, we employ the point-slope form to produce the equation for this perpendicular line.

$$y = -1/3 x - 8$$

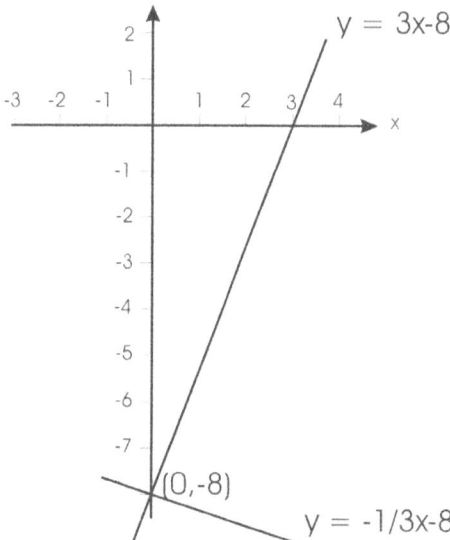

45

Appendix F

Equations, Functions
&
The Quadratic Formula

Can you distinguish between an equation and a function? Let us begin with an equation:
We will convert it into a function.

$$x^3 + 3x = 5x + 3$$ Solve the equation for zero (get zero on one side by itself)

$$x^3 - 2x - 3 = 0$$ Now, replace the zero with a functional notation

$$f(x) = x^3 - 2x - 3$$ Basically, all the terms are the same except that the zero has been replaced by a functional notation. Perhaps it is clear that the roots are the x-values at which the function equals zero.

We can employ the quadratic formula to find the roots of a 2nd degree polynomial function. In accordance with the Fundamental Theorem of Algebra, we expect to find two roots because two is the highest degree of any term in the expression. The roots of our creation are of our own choosing. Let us select 2 and 3 as our two roots. At the left, we will produce an equation and perform a slight adjustment to create a polynomial function. At the right, we will substitute these coefficients into the quadratic formula.

$(x-3)(x-2) = 0$ the roots are 3 and 2 $x = [-b \pm \sqrt{b^2 - 4ac}] / 2a$

$x^2 - 2x - 3x + 6 = 0$ after expanding $x = [5 \pm \sqrt{25 - 4(1)(6)}] / 2(1)$

$0 = x^2 - 5x + 6$ after 180 degree rotation

but keeping term sequence $x = [5 \pm 1] / 2$

$f(x) = x^2 - 5x + 6$ functional notation $x = 6/2, 4/2$

is substituted $x = 3, 2$ as we expected

We know that these are the correct roots because we created the original equation as such. Let's create another polynomial function of degree 2 so we can again check the validity of the quadratic formula.

$(x-4)(x-2) = 0$ we chose roots of 4 and 2 $x = [-b \pm \sqrt{b^2 - 4ac}] / 2a$

$x^2 - 6x + 8 = 0$ expand and simplify

$g(x) = x^2 - 6x + 8$ now, substitute $\rightarrow$ $x = [6 \pm \sqrt{36 - 4(1)(8)}] / 2(1)$

$x = 8/2, 4/2$

$x = 4, 2$ as we expected

It appears that we can have confidence in the quadratic formula. Have no fear! The world's greatest mathematicians assure us of this. In some cases, employing the quadratic formula is not necessary. Solving by factoring may be the easier method. Another method of solving a quadratic is by completing the square which is the manner in which the quadratic formula was derived. When completing the square, don't forget to divide the equation by the coefficient of the x-squared term immediately, before you forget. Next, get the constant term alone on the right side of the equation. Then divide the coefficient of the x-term by 2, square the result and add this to both sides of the equation. At this point, the left side of the equation should be a perfect square. Conclude by taking the square root of both sides.

Appendix G

The Vertical Line Test
&
Imaginary Roots

If we graph an equation, we can perform the vertical line test to determine whether or not we are working with a function. We simply drop a vertical line at a number of points to assure us that no vertical line intercepts the curve at more than one point. In other words, each x-value must have only one ordinate.

Also, we may expect the graph of every quadratic function to reveal two unique x-axis intercepts. But the great <u>mathematicians</u> of the late middle ages <u>were</u> stunned by the discovery of roots which did not intercept the x-axis. The root, $\sqrt{-1}$, does not appear on the number line of either rational or irrational numbers. In other words, it does not exist in the set of real numbers. Rene Descartes, who gave us the Cartesian coordinates, intensely disliked these strange numbers. He considered these numbers to be even worse than the hated negative numbers. He assigned a scornful name to these numbers and the name "imaginary numbers" stuck. The symbol, i, was employed as shorthand notation to identify such numbers.

To illustrate the above problem, let's graph two important parabolas.

Consider two nearly identical equations $i = \sqrt{-1}$

$y = x^2 - 1$ $y = x^2 + 1$

$0 = x^2 - 1$ set y = 0 to get the x-intercepts $0 = x^2 + 1$

$0 = (x - 1)(x + 1)$ factor both $0 = (x - i)(x + i)$

The roots are +1 or -1   The roots are +i or −i

The graph displays the two x-intercepts   No intercepts on an axis of Real numbers

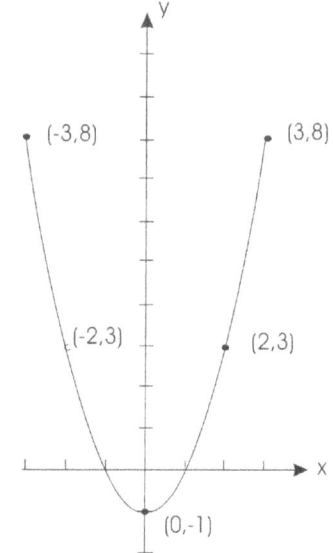
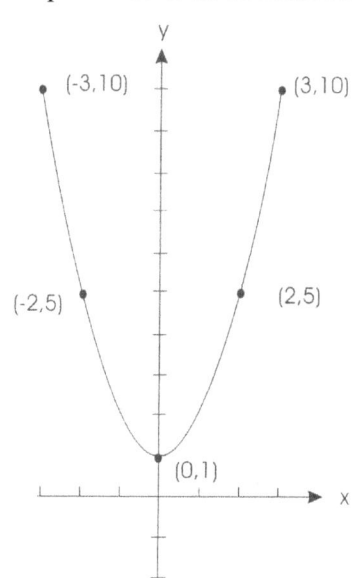

47

Appendix H
Recall Names Upon Sight

y = x
**linear
function**

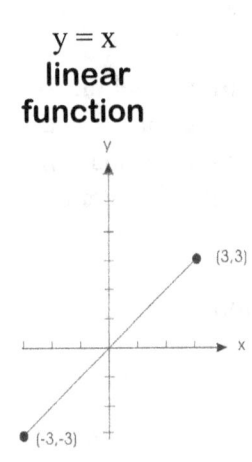

y = |x|
**absolute value
function**

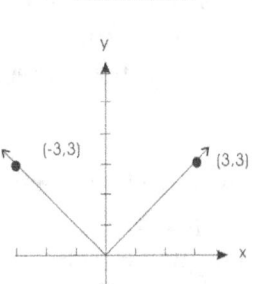

y = √x̄
**square root
function**

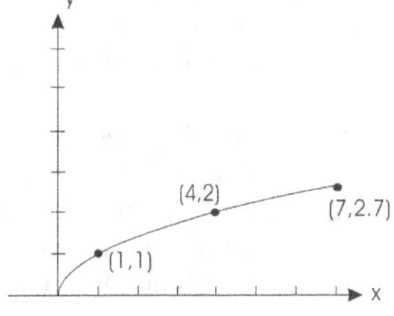

y = 1 / x
**hyperbolic
function**

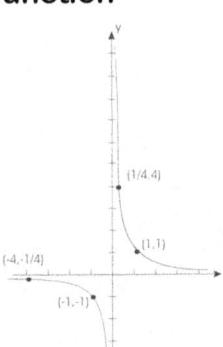

y = x²
**parabolic
function**

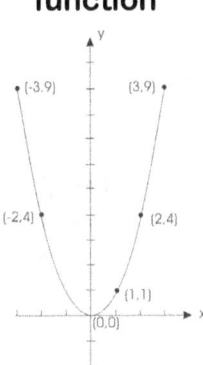

y = x³
**cubic
function**

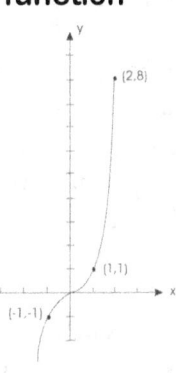

y = 2ˣ
**exponential
function**

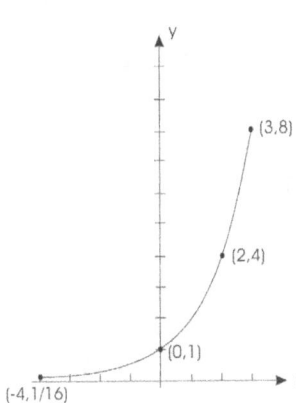

y = log₂ x
**logarithmic
function**

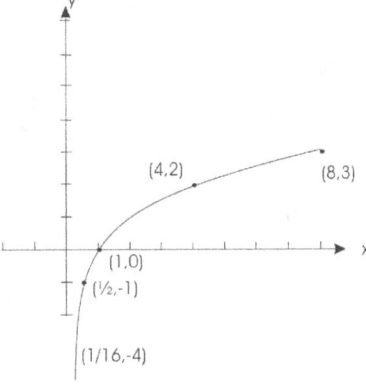

y = x⁴
**quartic
function**

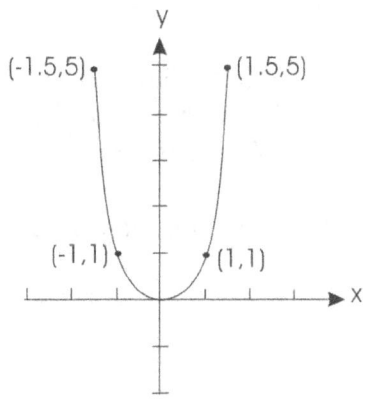

48

Appendix I

Managing the Power Formula

The power formula in Physics is chosen to illustrate a common problem in algebra. When presented with the formula, P = fs / t, suppose the student is asked to solve for t.   The normal student reaction is to begin substituting the known quantities into the equation. This tends to complicate the problem.   Normally, the student should solve the literal equation for the desired unknown.

Recall the lessons of Appendix A. There were four statements to memorize.
1.   Equals added to equals give equals.
2.   Equals subtracted from equals give equals.
3.   Equals multiplied by equals give equals.
4.   Equals divided by equals give equals.

In the form presented, the power equation is solved for P. We have three more forms of this equation. These forms are: 1. Solved for t, 2. Solved for f and 3. Solved for s. The student must be able to carry out these three solutions for the literal equation. We begin with the formula, P = fs / t.

To clear fractions, multiply both sides by t	$tP = t\,fs/\,t$
Associate the two t's on the right side	$tP = t/t\ fs$
$a/a = 1$ if $a \neq 0$, $1a = a$	$tP =\ 1\ fs$
Divide both sides by P	$t\,(P/P) = fs\,/\,P$
$a/a = 1$ if $a \neq 0$	$t\,(1)\ = fs\,/\,P$
$1(a) = a$	$t = fs\,/\,P$

Solve the equation for f	
Start with $tP = f\,s$ from above	$tP = fs$
Divide both sides by s	$tP/s = f\ s/s$
$a\,/\,a = 1$ if $a \neq 0$, $(1)a = a$	$tP/s =\ f\,(1)$
rotate 180 degrees, $(1)\,a =\ a$	$f = tP/s$
but maintain sequence of terms	

Solve the equation for s	
Start with $tP = fs$ from above	$tP = fs$
Divide both sides by f	$tP/f = f\,/f\,s$
$a\,/\,a = 1$ if $a \neq 0$	$tP/\,f = 1\ s$
$1\,a = a$, rotate 180 degrees	$s = tP\,/\,f$
but maintain sequence of terms	

Appendix J   Areas & Volumes

Square	Rectangle	Parallelogram	Trapezoid
	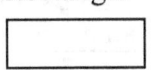	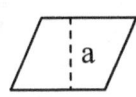	
$A = a\,b$	$A = a\,b$	$A = a\,b$	$A = a\,(b_1 + b_2) / 2$
altitude * base	altitude * base	altitude * base	altitude * (sum of bases) / 2

Circle	Ellipse	Triangle	Cube
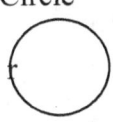	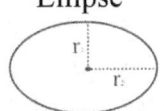		
$A = \pi\,r^2$	$A = \pi\,r_1\,r_2$	$A = \tfrac{1}{2}\,a\,b$	$A = 6\,l^2$
π r squared	π * product of radii	½ * altitude * base	6 * length squared

Volumes

Cube	Rectangular Prism	Cylinder	Pyramid
	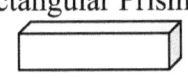	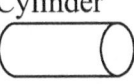	
$V = lwh$	$V = lwh$	$V = Bh$	$V = 1/3\,Bh$
base area * height	base area * height	base area * height	(base area * height) / 3

cone	sphere

$V = 1/3\,Bh$	$V = 4/3\,\pi\,r^3$
(base area * height) /3	4/3 π * (radius cubed)

50

Appendix K   Notes

symbol ------ name	symbol ------ name	explanation or comments
′ single prime	″ double prime	not quotation, denotes 1st or 2nd derivative
√ square root	π pi	root symbol, pi value is nearly 3.1416
Δ delta symbol	≤ less than or equal t	the Δ symbol denotes a change in value
∫ integral symbol	∞ infinity	sum, ∞ is greater than any real number
[left bracket	] right bracket	both employed to group terms
{ left brace	} right brace	both employed to group terms
⌈ left ceiling	⌉ right ceiling	both employed to group terms
≈ approximately	⇒ implies	nearly, implies is a guarantee
e base of natural logs	i=√-1	e value is about 2.7183, i is not a Real no.
dx horizontal differenti	dy vertical differentia	smaller than any named number > 0
‴ triple prime	> greater than	denotes the 3rd derivative, exceeds

The mathematical symbols were downloaded from Microsoft software to Word 2003. The less than, <, and greater than, >, symbols are available on the Microsoft keyboard

Appendix L   Phone Numbers

www.ingramcontent.com/pod-product-compliance
Lightning Source LLC
Chambersburg PA
CBHW081224170526
45165CB00009B/2935